Acta Biotechnologica

Volume 4 · 1984 · Number 3

Journal of microbial, biochemical and bioanalogous technology

Akademie-Verlag Berlin

ISSN 0138-4988
Acta Biotechnol., Berlin 4 (1984) 3, 193—304
EVP 30,— M

Instructions to Authors

1. Only original papers that have not previously been published will be accepted.
   Manuscripts may be submitted in English, German or Russian but always in duplicate. The name of the institute (inclusive full address) from which the manuscript originates should be stated below the author's name(s). The authors are responsible for the contents of their contributions.

2. Original papers should not consist of more than 20 typewritten pages (2 lined-spacing), including references, tables and figures; short original communications may have a maximum of 6 typewritten pages.

3. Each paper should be preceded by a summary one in English.

4. Latin names of species as well as passages to be printed in italics for greater emphasis should be signed with waved line. Please use only units and symbols of the SI-system.

5. Tables may be used to effect a reduction of the text or to make assimilation easier. They should be numbered consecutively throughout and supplied with a brief heading. They shouldn't be written among the text but on separate sheets.

6. Number and size of illustrations should be limited to the essential minimum, they should be numbered consecutively and brought on separate sheets.
   Line drawings including graphs and diagrams should be drawn in black ink. Half-tone illustrations should be supplied as white glossy prints. Figure legends are to be typed in sequence on a separate sheet.
   Authors name is to be written on to the rear of each sheet.

7. References, listed at the end of the contribution, should contain only works cited in the text. They should be numbered in the order in which they are at first mentioned in the text. Please give surnames and initials of all authors, name of journal abbreviated according to "Chemical Abstracts — List of Periodicals", volume number, year of publication, issue number or month, first page number; books are to be cited with full title, edition, volume number, page number, place of publication, publisher and year of publication.

8. Notes on the text may be given as footnotes on the relevant page.

9. 50 reprints are free of charge.
   Additional reprints may be ordered.

10. The author receives 2 galley proofs for correction.
    They are sent back to the redaction
    (Dr. Dimter, DDR - 7050 Leipzig, Permoserstr. 15)
    as soon as possible.

Acta Biotechnologica

Journal of microbial, biochemical and bioanalogous technology

Edited at
Institute of Technical Chemistry of the Academy of Sciences of the G.D.R.; Leipzig
and
Institute of Technical Microbiology; Berlin
by
M. Ringpfeil, Leipzig and G. Vetterlein, Berlin

Volume 4

1984

Number 3

AKADEMIE-VERLAG · BERLIN

"Acta Biotechnologica" publishes reviews, original papers, short communications and reports out of the whole area of biotechnology.
The journal shall promote the foundation of biotechnology as a new, homogeneous scientific field. According to biotechnology are microbial technology, biochemical technology and technology of synthesyzing and applying of bioanalogous reaction systems.
The technological character of the journal is guaranteed thereby that microbial, biochemical, chemical and physical contributions must show definitely the technological relation.

Terms of subscription for the journal "Acta Biotechnologica"
Orders can be sent

— **in the GDR:** to Postzeitungsvertrieb, or to Akademie-Verlag, DDR - 1086 Berlin, Leipziger Str. 3—4, PF-Nr. 1233;

— **in the other socialist countries:** to a book-shop for foreign language literature or to the competent news-distributing agency;

— **in the FRG and Berlin (West):** to a book-shop or to the wholesale distributing agency Kunst und Wissen, Erich Bieber OHG, Wilhelmstr. 4—6, D - 7000 Stuttgart 1;

— **in the other Western European countries:** to Kunst und Wissen, Erich Bieber GmbH, Dufourstr. 51, CH - 8008 Zürich;

— **in other countries:** to the international book- and journal-selling trade, to Buchexport, Volkseigener Außenhandelsbetrieb der DDR, DDR - 7010 Leipzig, Postfach 160; or to the Akademie-Verlag, DDR - 1086 Berlin, Leipziger Str. 3—4.

Acta Biotechnologica

Herausgeber: Institut für technische Chemie der AdW,
DDR - 7050 Leipzig, Permoserstr. 15
(Direktor: Prof. Dr. Manfred Ringpfeil) und

Institut für Technische Mikrobiologie
DDR - 1017 Berlin; Alt-Stralau 62
(Direktor: Dipl. Ing. G. Vetterlein)

Verlag: Akademie-Verlag, DDR - 1086 Berlin, Leipziger Straße 3—4;
Fernruf: 2 23 62 21 und 2 23 62 29; Telex-Nr.: 114420; Bank: Staatsbank der DDR, Berlin, Konto-Nr.: 6836-26-20712.

Redaktion: Dr. Lothar Dimter (Chefredakteur),
Käthe Geyler (Redakteur),
DDR - 7050 Leipzig, Permoserstr. 15; Tel.: 2 39 22 55.

Veröffentlicht unter der Lizenznummer 1671 des Presseamtes beim Vorsitzenden des Ministerrates der Deutschen Demokratischen Republik.

Gesamtherstellung: VEB Druckhaus „Maxim Gorki", DDR - 7400 Altenburg.

Erscheinungsweise: Die Zeitschrift „Acta Biotechnologica" erscheint jährlich in einem Band mit 4 Heften. Bezugspreis eines Bandes 120,— M zuzüglich Versandspesen; Preis je Heft 30,— M. Bestellnummer dieses Heftes: 1094/4/3.

AN (EDV) 42133

Acta Biotechnol. 4 (1984) 3, 195—208

Characteristics of Genetic Material with Reference to Patent Documents*

S. KNUTH[1], L. G. BELYAKOVA[2], V. V. VELKOV[2], V. N. DEMENTYEV[3], H. J. LOMMI[4], R. A. L. HARPER[1] and H. G. GYLLENBERG[1]

[1] Department of Microbiology, University of Helsinki, 00710 Helsinki 71, Finland
[2] Institute of Biochemistry and Physiology of Microorganisms USSR Academy of Science, Pushchino, Moscow region, 142292 USSR
[3] USSR State Committee of the Council of Ministers of the USSR on Inventions and Discoveries, M. Cherkassy per. 2/6, Moscow, USSR
[4] National Board of Patents and Registration Bulevardi 21, 00180 Helsinki 18, Finland

Summary

Characteristics of genetic material: plasmids, vectors, recombinant DNA and DNA fragments were, studied from 235 patent documents. These documents were from the patent offices of the Federal, Republic of Germany (38), the German Democratic Republic (15), the European Patent Office (53), Finland (43), Great Britain (65), the Soviet Union (3), and the United States of America (18). Attention was paid also to the terminology used in the patent documents. According to this study the way in which the genetic material was characterized in patent documents of different countries was quite variable, and the terminology used was sometimes confusing.

Introduction

Thirty years ago WATSON and CRICK published their paper on the structure of DNA and so opened a new era where gene expression and regulation could be studied at a molecular level [15]. It was not until BOYER and COHEN developed the techniques for producing recombinant DNA that the potential of this powerful technique in genetics was realised [16]. Today, microorganisms that have been engineered, are being used for the production of valuable compounds such as interferon, insulin, and human growth hormone.

In 1980 the Supreme Court in the USA regarded as protectable subject matter a strain of *Pseudomonas* which had "at least two plasmids" from other hosts inserted into it [6]. CHAKRABARTY was granted a patent for this invention. The court ruled that the distinction between "living" and "non-living" matter should not be an obstacle for granting patents. Following this precedent setting decision the number of granted US-patents concerning engineered organisms increased. According to SUETINA et al. [11] year 1981 (as compared to the previous year) saw a worldwide increase of 100% in the number of patent applications that involved recombinant DNA techniques.

Although the rapid development in the field of molecular biology and the practical utility of the results have given rise to a dramatic increase in the number of patent

* This study is part of a cooperative project between the Academy of Finland, and the USSR Academy of Science. The participating institutions are the Department of Microbiology, University of Helsinki (Finland), the National Board of Patents and Registration (Finland), the Institute of Biochemistry and Physiology of Microorganisms, Pushchino (USSR), and the State Committee for Inventions and Discoveries (USSR).

applications in this field, the number of patents that have been granted remains small in comparison to the number of stillpending patent applications. One reason for this is the fact that molecular biology is a new field for patent officials, and the examination of the applications requires that they familiarise themselves with the terminology used in this branch of science. Progress has also been hampered, because at the moment there are no rules or recommendations in the patent legislation of the different countries which would help scientists to describe their inventions in the field of genetic engineering. It is therefore obvious that both patent officials and scientists would greatly benefit if clear guidelines or at least recommendations were established.

In many countries inventions are patentable if they fulfill the following criteria, namely, the invention must be novel, non-obvious, and can be used industrially, which includes for example use in agriculture. It must also be reproducible by anyone "skilled in the art" [4]. Many problems arise when it has to be decided if an invention is industrially useful, or if sufficient information has been given in the patent application to allow the invention to be reproduced. From the point of view of patent protection the scientist would like as broad a protection as possible, yet he does not know how the application should be formulated so that it will receive favourable consideration from the patent office. These problems are especially relevant in the field of gene technology. At one extreme the DNA molecule may be claimed in terms of its base sequence, while at the other extreme the same molecule is claimed as a combination of its functional sub-units [2]. Part of the problem exists because there are as yet too few precedents to which reference can be made, and doubtless, only with the passage of time, will experience be gained in the handling of patent applications dealing with genetic material.

Inventions in the field of genetic engineering have been discussed in several recent papers [1, 4, 6, 7, 8, 12, 14]. However only a few papers deal with the characterization of novel genetic material. HALLUIN [6] has presented a list of characteristics that could be used in combination with deposition for defining microorganisms, plasmids or vectors that are newly isolated, novel or man-made. The suggestions are as follows.

- Characterization of the microorganism by genus and species ... in case of a plasmid or vector, the accepted classical name ...
- A detailed description of how the organism, plasmid or vector was obtained .... how it was modified (mutagenesis or recombinant techniques)
- A morphological description of the organism ...
- Metabolic characteristics of the organism ...
- Effect of various external factors on the organism, plasmid or vector such as resistance to antibiotics, restriction enzyme sites in the case of plasmids ...
- Suitable growth conditions ...
- Description of how to put the organism into use ...
- A description of the utility of the invention ...

SALIWANCHICK [10] has given nine "forms" of disclosure which could be used for characterizing vectors, DNA sequences, and genes. A vector if it is a novel plasmid or bacteriophage should be characterized by its restriction enzyme cleavage map. A DNA sequence should be specified by its source and nucleotide sequence, and for a gene all control and regulatory regions surrounding the gene should be given.

The purpose of this study has been to analyse the terms that are used in defining genetic material in patent applications, and granted patents with the aim that a list of essential features could be drawn up that would be helpful to both patent officials and scientists. The scope of the study has encompassed the patent documents of the patent offices from many different countries, and the features used to characterize plasmids, vector, recombinant DNA, and DNA fragments have been compared. Special attention has been paid to the terminology that is used in patent documents and to the extent of the claims and specifications.

Materials and Methods

Patent documents from the patent offices of the Federal Republic of Germany (DE), the German Democratic Republic (DD), the European Patent Office (EP), Finland (FI), Great Britain (GB), Sweden, Norway, and Denmark (SC), the Soviet Union (SU), and the United States of America (US) were studied. The work was divided (Table 1) so that the researchers from the Soviet Union studied the patent documents from DD, DE, EP, and SU. Before January 1st 1982 the following patent documents had been studied 15 DD, 38 DE, 53 EP and 3 SU. In Finland the researchers dealt with the patents from FI, GB, SC, and US, and by August 1st 1982, had reviewed 43 FI, 65 GB, and 18 US patents or patent applications. The Scandinavian patent documents (SC) were only available as abstracts, and since the information contained therein was so small, it could not be used for comparison.

Patent documents consist essentially of two parts, a descriptive part known as the specification, and the claims part including the description of subject(s) claimed, drawn up in conformity with the legal rules. The Soviet researchers dealt with both claims and specifications, whereas the Finnish group dealt only with claims. In addition the Finnish group made the distinction between patent applications and granted patents. Patent applications refer to documents that are under examination, while a granted patent means that the patent document has gone through the process of inspection and now affords protection to the invention. It should be noted that the US patent office normally publishes only information on patents that have been granted, and a similar system exists for Soviet patents and inventors certificates, but other patent offices publish patent applications already 18 months after they have been filed for the first time.

Table 1. Allocation of patent documents

	Finnish group	Soviet group
Documents studied	FI, GB, US[1]	DD, DE, EP, SU[1]
Deadline	August 1982	January 1982
Scope	Claims	Claims & Specifications
Distinction	Separation of granted patents from applications	No distinction
Total number	126	109
Total inventions	73	71

[1] US documents deal only with granted patents. SU documents deal with granted patents or inventors certificates.

Differences existed in the listing of the features that were considered essential in defining plasmids, vectors, recombinant DNA, and DNA fragments, and attempts have been made to arrange the material in such a way that a direct comparison could be made.

The contributions of the Finnish and Soviet research groups are indicated in Table 1.

The documents under discussion are mostly classified in C12N 15/00 ("mutation or genetic engineering") according to the International Patent Classification 3rd edition. In patent documents there are claims for 1) plasmids, 2) vectors, 3) recombinant DNA, 4) DNA fragments (genes, adaptors and linkers) 5) microorganisms, and 6) phages, not only as "processes" for obtaining them, but also as "products". In the DD and SU patent

documents there were no claims for plasmids, vectors, recombinant DNA, or DNA fragments as products.
The number of documents with claims for novel genetic material which includes plasmids, vectors, recombinant DNA, and DNA fragments as a product can be seen in Table 2.
For the sake of clarity and convenience plasmids, vectors, and recombinant DNA were grouped together in Tables, while DNA fragments were dealt with separately (Tables 3—5). The tables show the number of documents with subject features and the percentage of such documents studied. A survey of the material showed that there were 12

Table 2. Patentable genetic material

Patentable genetic material	EP	DE	FI	GB	US
1) Plasmid	26	20	9	28	6
2) Vector	21	10	16	15	1
3) Recombinant DNA	8	3	7	10	0
4) DNA fragment	14	6	8	18	1

primary characteristics which were to a great extent common for plasmids, vectors, and recombinant DNA, and DNA fragments. There were also 7 secondary characteristics which were not used so often, but in some cases they could be essential, depending on the invention. The list of primary characters used to define genetic material is as follows

— Name (designation) (name of plasmid or vector e.g. pUC 6)
— Constituent parts (functional sub-units)
— Name of specific structural gene
  (or encoded product)
— Restriction enzyme cleavage sites of plasmid, vector, or recombinant DNA
  Restriction enzyme cleavage sites of DNA insert.
— Molecular weight/length of plasmid, vector, or recombinant DNA.
  Molecular weight/length of DNA insert.
— DNA produced or isolated by specific process
— Host range (species or genus or group of suitable hosts)
— Function (transfer, cloning, or expression)
— Order of genes (localization of genes)
— Origin of DNA insert (isolated, chemically synthesised, cDNA)
— Nucleotide sequence
— -complete (complete base sequence)
— -partial (only terminal nucleotides of DNA fragment)
— -with wide range of variations
— Regulatory region
— -promoter
— -operator
— -ribosome binding site
— -terminator
— -signal sequence

The list of secondary characteristics is as follows

— Physical map (scheme of positions of restriction sites)
— Genetic markers (e.g. antibiotic resistance)
— Adaptor, linker, or other DNA fragments
— Order (localization) of DNA fragments other than gene.
— Ability to hybridize with specific RNA or DNA.
— Genetic map (scheme of positions of genes)
— Copy number per cell.

Results

Each patent office has to follow and interpret the national patent law, yet it would seem that there are no big differences in the terms used for the purposes of characterization. This may be due to the fact that the majority of the patent documents studied so far exist only as applications, and their final and corrected form has not yet been decided. The US and the SU are exceptions in that their patent documents are all granted patents. The number of patents that have been granted remains small, so unfortunately a meaningful comparison between patent applications and granted patents could not be done. Even though the patent documents from FI, GB, and US have been studied seven months longer than the DD, DE and EP and SU documents, the results show that with few exceptions there is good agreement in the terms used to characterize genetic material The exceptions (results of the Finnish versus the Soviet group) include the physical map of a plasmid (34.1%/18.4%), the constituent parts of a vector (92.9%/40.7%), the function of a vector (75.0%/48.1%), and the constituent parts of recombinant DNA (100%/ 54.5%). These differences may exist because scientists do not know what features are the most important from a legal point of view.

Table 3 summarises the material studied by the Finnish group. It indicates that plasmids are most frequently characterized by name and constituent parts. Vectors are characterized by constituent parts, name of specific structural gene, function, and regulatory region. In this material recombinant DNA is most frequently characterized by name, constituent parts, name of specific structural gene, DNA produced by a specific process, and regulatory region.

The characterization of patentable material consisting of plasmids, vectors, and recombinant DNA, and studied by the Soviet group is presented in Table 4. Plasmids are most frequently characterized by name and constituent parts. For vectors the most frequent terms used are constituent parts, name of specific structural gene, function, and regulatory region, respectively. As for recombinant DNA the same set of features are most frequently used.

Table 5 presents the material studied by both groups concerning the characterization of DNA fragments. There is a very clear correspondence concerning name of specific structural gene and the origin of the DNA fragment.

Discussion

The way in which genetic material was characterized in the patent documents of different countries was quite variable, and the terminology used was sometimes confusing. The meaning of the terms varied and they were quite often used in a way that would not normally be encountered in research publications. The following text sums up the problems that exist at the moment.

The term DNA fragment in this study covers expressions such as gene, DNA sequence, DNA molecule, polynucleotide, nucleic acid, cDNA, adaptor, or linker. In patent documents a DNA fragment usually refers to a specific DNA sequence which codes for a specific product (a structural gene). If a regulatory region is attached to a specific DNA sequence, then in some patent documents the genetic material is still referred to as DNA fragment, while in other documents it would be called recombinant DNA. Obviously there is confusion in the terminology. It should be stated that both terms are correct, and it is only a matter of fancy into which category the scientist places his invention. In general, however, in most patent applications, recombinant DNA consists of a combination of regulatory region, a foreign DNA insert, packaged into a vector, (plasmid or phage).

Table 3. Characterization of patentable genetic material (Finnish group)

Publication country	Plasmid				
	FI	GB	US	Pat. appl. & Pat[1]	
Total document	9	29	6	41	
Features				Doc	%
1 Name (designation)	8	25	6	36	87.8***
2 Constituent parts	3	14	1	17	41.5**
3 Name of specific structural gene (or encoded product)	2	8		9	22.0*
4 Restriction enzyme cleavage sites	2	9	1	11	26.8*
— (DNA insert)		1	1	2	4.9*
5 Molecular weight/lenght	1	9	1	11	26.8*
— (DNA insert)		1	1	2	4.9*
6 DNA produced by specific process	1	3		4	9.8*
7 Host range	2	7		8	19.5*
8 Function		4		4	9.8*
9 Order of genes	1	5		5	12.2*
10 Origin of (DNA insert)	2	6		7	17.1*
11 Nucleotide sequence of (DNA insert)		1	1	2	4.9*
— complete		1		1	2.4*
— partial			1	1	2.4*
— with wide range of variations					
12 Regulatory region	2	7		8	19.5*
— promoter	2	5		6	14.6*
— operator		2		2	4.9*
— ribosome binding site	1	1		2	4.9*
— terminator		2		2	4.9*
— signal sequence					
13 Physical map		9	5	14	34.1**
14 Genetic markers	1	5		5	12.2*
15 Adaptor, linker or other DNA fragment		1		1	2.4*
16 Order (localization) of other DNA fragments		1		1	2.4*
17 Ability of (DNA insert) to hybridize with specific RNA and DNA					
18 Genetic map					
19 Copy number per cell					

60/100% = *** 30/60% = ** 0/30% = *

[1] Analogs are taken into account

Vector					Recombinant DNA				
FI 16	GB 15	US 1	Pat. appl. & Pat[1] 28		FI 7	GB 10	US 0	Pat. appl. & Pat[1] 14	
			Doc	%				Doc	%
2	6		7	25.0*	5	4		7	50.0**
15	14	1	26	92.9***	7	10		14	100.0***
12	8	1	19	67.9***	6	7		10	71.4***
2	3		5	17.9*	2			2	14.3*
1	1		2	7.1*		1		1	7.1*
1	2		3	10.7*					
	4		4	14.3*	2	5		6	42.9**
6	4		8	28.6*	2	3		3	21.4*
15	9	1	21	75.0**					
4	4		6	21.4*		2		2	14.3*
	5	1	6	21.4*	1	3		3	21.4*
1	2	1	4	14.3*	4	2		4	28.6*
1	2	1	4	14.3*	4	2		4	28.6*
	2		2	7.1*					
7	7		11	39.3**	3	5		7	50.0**
6	6		10	35.7**	2	3		5	35.7**
4	3		6	21.4*	2			2	14.3*
5	4		7	25.0*		1		1	7.1*
1	1		1	3.6*		1		1	7.1*
2			2	7.1*	3	2		3	21.4*
					1			1	7.1*
	1		1	3.6*					

Table 4. Characterization of patentable genetic material (Soviet group)

Publication Country Total documents	Plasmid					
	EP 18		DE 20		Total[1] 38	
(Cl. = Claims SP. = Specification)	CL.	SP.	CL.	SP.	CL. %	SP. %
1 Name (designation)	11	2	13	1	63.1***	7.9*
2 Constituent parts	9	4	5	10	56.8**	36.8**
3 Name of specific structural gene (or encoded product)	6	2	6	5	31.6**	18.4*
4 Restriction enzyme cleavage sites	3	10	4	5	18.4*	39.5**
— (DNA insert)	1		2		7.9*	
5 Molecular weight/length	2	3	8	2	26.3*	13.2*
— (DNA insert)	1	1	1	1	5.3*	5.3*
6 DNA produced by specific process	1				2.6*	
7 Host range	3	12	3	14	15.8*	68.4**
8 Function		6	2	7	5.3*	34.2**
9 Order of genes	1		3		10.5*	
10 Origin — (DNA insert)	3		1		10.5*	
11 Nucleotide sequence of (DNA insert)		3	1	1	2.6*	10.5*
— complete		1	1	1	2.6*	5.3*
— partial		2		1		7.9*
— with wide range of variations			1		2.6*	
12 Regulatory region		2	4		10.5*	5.3*
— promoter		2	4		10.5*	5.3*
— operator		2	2		5.3*	5.3*
— ribosome binding site				1		2.6*
— terminator						
— signal sequence						
13 Physical map	2	8	5	7	18.4*	39.5**
14 Genetic markers	2	3	3	2	13.1*	13.1*
15 Adaptor, linker or other DNA fragment						
16 Order (localization) of other DNA fragments						
17 Ability of (DNA insert) to hybridize with specific RNA and DNA						
18 Genetic map		2		8		26.3*
19 Copy number per cell			2	5	5.3*	13.1*

60/100% = *** 30/60% = ** 0/30% = *

[1] Analogs are taken into account.

Vector						Recombinant DNA					
EP 17		DE 10		Total[1] 27		EP 8		DE 3		Total[1] 11	
CL.	SP.	CL.	SP.	CL. %	SP. %	CL.	SP.	CL.	SP.	CL. %	SP. %
5	3	3	2	29.6*	18.5*	3				27.3*	
7	3	4	1	40.7**	14.8*	4		2		54.5**	
9	2	8		63.0***	7.4*	6		2		72.7***	
4	4	1	2	18.5*	22.2*	1	3	1		18.2*	27.3*
2				7.4*		1		1		18.2*	
	6	1	3	3.7*	33.3**	1	3			9.1*	27.3*
1	3		1	3.7*	14.8*	1	2			9.1*	18.2*
		1		3.7*		1				9.1*	
5	4	1	2	22.2*	22.2*	2	5		1	18.2*	54.5**
11	1	2		48.1**	3.7*	2				18.2*	
5		1		22.2*		2	1			18.2*	9.1*
						3				27.3*	
2	11	3	2	18.5*	48.1**	2			1	18.2*	9.1*
2	4	2	3	14.8*	25.9*	1			1	9.1*	9.1*
	9	1		3.7*	33.3**	1				9.1*	
4		3		25.9*		4				36.4**	
7	1	2	2	33.3**	11.1*	6		1		63.6***	
6	1	2	2	29.6*	11.1*	5		1		54.5**	
5	1	1	1	22.2*	7.4*	3		1		36.4**	
3				11.1*			1				9.1*
1				3.7*							
2				7.4*		1				9.1*	
	13		6		70.4***		6		2		72.7***
1	1			3.7*	3.7*	1	1	2		27.3*	9.1*
	1		2		11.1*	1			1	9.1*	9.1*
4		1		18.5*				1		9.1*	
		1		3.7*							
	10		5		55.5**		4		2		54.5**
							1				9.1*

Table 5. Characterisation of patentable genetic material (DNA Fragment)

Publication country	Finnish group					Soviet group					
Total documents	FI 8	GB 18	US 1	Pat. appl. & Pat[1] 24		EP 13		DE 6		Total[1] 19	
(Cl. = Claims SP. = Specification) Features				Doc	%	CL.	SP.	CL.	SP.	CL. %	SP. %
1 Name (designation)		1		1	4.2*	2				10.5*	
2 Constituent parts											
3 Name of specific structural gene (encoded product or function)	8	15	1	21	87.5***	11		6		89.5***	
4 Restriction enzyme cleavage sites	3	4	1	7	29.2*	3	3	1		21.1*	15.8*
5 Molecular weight/length	1	1	1	3	12.5*	2	2		1	10.5*	15.8*
6 DNA produced by a specific process	1	8		8	33.3**						
7 Host range	1	1		1	4.2*		6		1		36.8**
8 Function											
9 Order (localization of genes)		3		3	12.5*	1		1		10.5*	
10 Origin	2	8		10	41.7**	6		3		47.3**	
11 Nucleotide sequence	3	5	1	8	33.3**	6	3	1		36.9**	15.8*
— complete	3	4	1	7	29.2*	4	3			21.1*	15.8*
— partial		1		1	4.2*	3				15.8*	
— with a wide range of variations		1		1	4.2*	5		1		31.6**	
12 Regulatory region	2	5		6	25.0*	3				15.8*	
— promoter		4		4	16.7*	1				5.3*	
— operator									1		5.3*
— ribosome binding site		3		3	12.5*			1		5.3*	
— terminator											
— signal sequence	4	1		4	16.7*	1				5.3*	

13 Physical map							6				31.6**
14 Genetic markers											
15 Adaptor, linker, or other DNA fragments		2	1	3	12.5*	1	2		1	5.3*	15.8*
16 Order (localization) of other DNA fragments		2		2	8.3*						
17 Ability of insert to hybridize with specific RNA or DNA	2	2		4	16.7*	3	1			15.8*	5.3*
18 Genetic map							2				10.5*
19 Copy number per cell											

60/100% = ***   30/60% = **   0/30% = *

[1] Analogs are taken into account.

The term vector also had various meanings in patent documents. Most often it could be regarded as a plasmid or a phage to which a foreign DNA insert has been linked. But it could also mean that it is a vehicle especially intended for gene cloning, and not yet containing any foreign DNA insert.
For plasmids, patent protection has been claimed for both natural plasmids and also for artificially constructed recombinant plasmids. There was usually no indication in the patent claims as to whether a plasmid was natural or artificially constructed, though from the specifications this differentiation becomes apparent.
The terms used to characterize plasmids and vectors were in many cases very similar in both patent applications and granted patents. From a historical point of view, in the first patents that were granted for plasmids and vectors, the features listed were rather limited, and gave very broad protection to the invention. For example in the US 4322 499 patent a claim was approved for a plasmid which contained a foreign DNA fragment (cDNA for the endorphin gene) and the only given characteristic in the claims was the designation pBR322/ME-150. Also in patent specification GB 1568047 there was a claim for a recombinant transfer vector which comprised a plasmid and codons for a specific product (human chorionic somatomammotropin). Designation of the plasmid and a wide range of nucleotide sequences were presented. It is now apparent that the number of details has increased in the latest patent applications and the information about the regulatory region and its location, location of structural genes, or other DNA inserts have now been included. It could be surmised that this increase in data reflects the technical developments in this fast growing field of science.
It has to be borne in mind that plasmids or vectors are very rarely claimed alone in patent documents, but appear in a combination e.g. as, a) process for producing a plasmid, b) plasmid as a product, c) vector, d) process for transforming a microorganism, e) transformed microorganism as a product.
One important point which also needs to be taken into consideration is the question of deposition. It is difficult if not impossible to write a patent specification that would enable someone "skilled in the art" to obtain or "create" the strain for himself, since the organism may not be easy to locate, isolate, or produce. The practice has therefore arisen of depositing microorganisms in culture collections, so that they would be available to third parties under a resonable set of conditions [5, 13].
The deposition of plasmids, vectors, or recombinant DNA has also been the subject of debate. Some would argue that a detailed description is quite sufficient for someone "skilled in the art" to prepare the required material. The argument has been that a molecule of DNA is much easier to define in chemical terms than the vastly more complicated microorganism [2]. While on the other hand some say that since the genetic code is by nature degenerative, the desired product may not be obtained even though the procedures in the specification have been followed exactly [6]. This then raises the interesting question of how would patent offices view a sequence that was substantially different from existing patented sequences, yet gives better results [3, 9]. There is also the view that organisms that have been genetically engineered, plasmids and cloned genes should all be deposited [10].
In the patent documents under study, the deposition of genetic material was rarely presented in the claims, more often it was presented in the specifications. Deposition of recipient microorganisms and transformed host microorganisms could be found most often. Plasmids, vectors and recombinant DNA were rarely deposited as a DNA preparation, but in slightly more than 50% of the cases were included within a host microorganism. The data on the deposition of genes was not available in any of the documents. It has been proposed that natural plasmids should be deposited within their host microorganism, but vectors that have resulted from recombinant techniques would only require a written description. If the deposit is not necessary, this is a benefit to the scientist

since his plasmid would not fall into the wrong hands. An interesting analogy has been drawn between DNA sequences and computer programmes, in that both store information that will later perform a certain task. The soft-ware however is not functional without being inserted into the hard-ware, and similarly DNA spliced in vitro is totally useless until it is loaded into a host cell which can replicate and express it [3]. This then is a strong argument against the deposition of recombinant DNA as such since it would be merely a programme without a computer.

Conclusions

In the patent documents studied the following characteristics were used most often in defining plasmids (in claims):
Name (designation) (87.8%/63.1%)
Constituent parts (41.5%/56.8%)
Physical map (34.1%/18.4%)
Name of specific structural gene or encoded product (22.0%/31.6%)
(additionally in specifications):
Host range, Restriction enzyme cleavage sites, Function.
For vectors the characteristics most often mentioned (in claims) were:
Constituent parts (92.9%/40.7%)
Function (75.0%/48.1%)
Name of specific structural gene or encoded product (67.9%/63.0%)
Regulatory region (39.3%/33.3%)
(additionally in specifications):
Physical map, Genetic map, Nucleotide sequence.
For recombinant DNA the following characteristics seem to be the most important (mentioned most often in claims):
Constituent parts (100%/54.5%)
Name of specific structural gene or encoded product (71.4%/72.7%)
Name (designation) (50.0%/27.3%)
Regulatory region (50.0%/63.6%)
Produced by a specific process (42.9%/9.1%)
(additionally in specifications):
Physical map, Host range, Genetic map.
For DNA fragment (gene) the following terms were used (in claims):
Name of specific structural gene or encoded product (87.5%/89.5%)
Origin (41.7%/47.3%)
Nucleotide sequence (33.3%/36.9%)
Produced by a specific process (33.3%/0)
(additionally in specifications):
Physical map, Host range.

The authors are greatly indebted to Dr. R. L. SUETINA for her tremendous work in searching for the patent documents used in the present study and to Academician G. K. SKRYABIN for his keen interest in and support of this work. We also wish to thank Ms. E. KOIVISTO for consultation and advice.

References

[1] BAUMBACH, F.: Der Neuerer **80 B** (1980) 10, 135.
[2] BENT, S. A.: J. Pat. Off. Soc. **64** (1982) 2, 60.
[3] BRENNER, S.: Recombinants that are the same but different. In: D. W. Plant, N. J. Reimers, and N. D. Zinder (Eds.) Banbury report 10. Patenting of life forms, pp. 43—47, Cold Spring Harbour Laboratory 1982.
[4] CRESPI, R. S.: Patenting in the biological sciences. 211p. John Wiley & Sons. Chichester. New York, 1982.
[5] GYLLENBERG, H.: Acta Biotechnol. **2** (1982) 3, 207.
[6] HALLUIN, A. P.: Patenting the results of genetic engineering research: An overview. In D. W. Plant, N. J. Reimers, and N. D. Zinder (Eds.) Banbury report 10. Patenting of life forms, pp. 67—126, Cold Spring Harbour Laboratory 1982.
[7] HÜNI, A. BUSS, V.: Ind. Prop. **21** (1982) 356.
[8] JACKSON, D. A.: Patenting of genes: What will the ground rules be? In: R. F. Acker and M. Schaechter (Eds.) Patentability of microorganisms: Issues and questions, pp. 23—27, American Society for Microbiology, 1981.
[9] KILEY, T. D.: Speculations on proprietary rights and biotechnology. In: D. W. Plant, N. J. Reimers, and N. D. Zinder (Eds.) Banbury report 10. Patenting of life forms, pp. 191—209, Cold Spring Harbour Laboratory 1982.
[10] SALIWANCHIK, R.: Legal protection for microbiological and genetic engineering inventions. 256p. London, Amsterdam: Addison-Wesley Publishing Company Inc. 1982.
[11] SUETINA, R. L., BELYAKOVA, L. G., VELKOV, V. V.: Biotechnology and recombinant DNA. 237p. Scientific Centre of Biological Research of the Academy of Sciences of the USSR in Pushchino, 1982.
[12] VOSSIUS, V.: Gew. Rechtschutz Urheberrecht **81** (1979), 579.
[13] VOSSIUS, V.: Patent protection for biological inventions. In H. J. Rehm and G. Reed (Eds.) Biotechnology Vol. 1. Microbial Fundamentals, pp. 435—452, Weinheim. Deerfield Beach Florida. Basel: Verlag Chemie 1981.
[14] VOSSIUS, V.: The Patenting of life forms under the European Patent Convention and German Patent Law: Patentable inventions in the field of genetic manipulations. In: D. W. Plant, N. J. Reimers, and N. D. Zinder (Eds.) Banbury report 10. Patenting of life forms, pp. 67 — 126, Cold Spring Harbour Laboratory 1982.
[15] WATSON, J. D.: Nature (London) **171** (1953) 737.
[16] WATSON, J. D.: Nature (London), **302** (1983), 651.

Acta Biotechnol. 4 (1984) 3, 209–234

Mikrocomputer-unterstützte Prozeßkontrolle in der Fermentationstechnik

K. Bayer

Institut für Angewandte Mikrobiologie
Universität für Bodenkultur
A-1190 Wien, Peter-Jordan-Straße 82, Austria

Summary

Nowadays computer control is an effective and well established tool in optimizing fermentation processes.
Based on recent development in microelectronics, especially in the field of personal computers, cost effective applications of this technology are now possible, even on laboratory- and pilot scale.
In this paper the application of a personal computer (CBM 3032), controlling various types of fermentations, is described.
In more detail the availability of physiologically relevant sensors, especially the respiration and the dynamic calorimetry, are discussed.
Finally some examples of practical realisations are presented.

Die Fermentationsindustrie hat sich in den letzten Jahren zu einem bedeutenden Wirtschaftsfaktor entwickelt. Allein in der Antibiotikaproduktion wurden beispielsweise im Jahre 1978 weltweit 10 Milliarden D-Mark Umsatz erzielt.
Auf Grund des ökonomischen Potentials der Fermentationsindustrie konzentriert sich die Forschung neben der Innovation neuer Verfahren auf die Analyse und Optimierung bereits bestehender, was durch Entwicklungen der Mikroelektronik massiv unterstützt wird.

1. Möglichkeiten der Prozeßoptimierung

Das physiologische Potential eines Mikroorganismus ist grundsätzlich genetisch determiniert. Inwieweit nun diese latenten Eigenschaften auch exprimiert werden, hängt wesentlich von den optimierten Umweltbedingungen ab. Dies kann durch eine geschickte Synthese von Theorie und experimentell fundierten Kenntnissen erzielt werden. Auf dieser Basis können Modelle aufgestellt werden, die einen realitätsbezogenen Einblick in die Komplexheit mikrobiologischer Prozesse geben können.

1.1. Modellbildung

Das Modell zeigt dann unter bestimmten Bedingungen annähernd das Verhalten jenes Teiles der Wirklichkeit oder zumindest jener Aspekte, die im Modell integriert sind.
Für die Computer-orientierte Prozeßkontrolle ist es unabdingbar, gewisse Verhaltensweisen der Mikroorganismen mathematisch formulieren zu können. Diese Algorithmen

können dann in die entsprechende Software zur Prozeßsteuerung und -regelung aufgenommen werden.
Eine Zusammenfassung der in der Fermentationstechnik angewendeten Modelle findet man bei BLANCH und DUNN [1] und bei ROELS und KOSSEN [2]. Eine repräsentative Auswahl von Modellen, die teils auch praktische Anwendungen enthalten, ist in Tabelle 1 zusammengefaßt. In der einschlägigen Literatur findet man vordergründig drei Prozesse, an denen Modellbildung betrieben wurde, und zwar die Belebtschlammtechnologie, die Produktion von Sekundärmetaboliten (Antibiotika) und in etwas geringerem Maße die Modellierung der Backhefetechnologie. In der Belebtschlammtechnologie versucht man verstärkt den „Kernprozeß" physiologisch in den Griff zu bekommen [19], [12]. Ausgehend von der MONOD-Kinetik — diese soll nur mit Vorbehalt angewendet werden — werden aus der Atmungsrate Modelle für den Substratabbau aufgestellt. Die Problematik besteht darin, daß die Substrataufnahme nicht der Verwertung proportional ist. Ähnliche Beobachtungen wurden auch an Hefe (*Candida intermedia*) im Labor des Autors gemacht [20], [21].
Ein weiteres Anwendungsfeld der Modellbildung ist die Backhefefermentation. Die Volumenzunahme erfolgt exponentiell, was durch einen Rechner leicht realisiert werden kann. Exaktere Modelle, wie sie von WANG et al. [22], PERINGER und BLACHERE [15] und SPRUYTENBERG et al. [23] aufgestellt wurden, berücksichtigen noch den physiologischen Status der Kultur (Atmungsrate) und regeln danach den Substratzulauf.
Zusammenfassend kann festgestellt werden, daß der mikrobiologische Prozeß durch die Modellbildung überschaubarer wird. Er kann unter verschiedensten Bedingungen simuliert werden. Verschiedene Computerfirmen bieten heute Simulationssprachen an, z. B. macht die von IBM einen Digitalcomputer in der Programmierung ähnlich einfach wie einen Analogcomputer [24].
Eine genaue Übersicht über den Einsatz von Simulationstechniken bei Fermentationen findet sich auch bei BLANCH und DUNN [1]. Neben der Simulation kann die Gültigkeit der getroffenen Annahmen statistisch überprüft werden. Empfindlichkeitsanalysen können durchgeführt werden.
Entsprechend dem heutigen Trend sollten vermehrt strukturierte Modelle eingesetzt werden. Dem steht aber meist die zu geringe Anzahl an meßbaren Parametern gegenüber.

1.2. Prozeßoptimierung

Analog zu obigen Ausführungen über die Modellbildung erfolgt die Optimierung eines Fermentationsprozesses. MUZYCHENKO et al. [25] versuchten auf theoretischer Basis [26] eine Prozeßoptimierung durchzuführen. Die optimalen Werte für die kontinuierliche Fermentation wurden aus batch-Kulturen (MONOD-Kinetik) ermittelt. Die Übergangszustände wurden mittels PONTRYAGIN's Maximum-Prinzip [27, 28] berechnet. Das Modell beinhaltet alle Schwächen der MONOD-Kinetik, es fehlt die experimentelle Verifizierung.
Häufiger wendet man Methoden an, die basierend auf jeweils experimentell ermittelten Daten die einzelnen Optimierungsschritte determinieren, womit meist auch die experimentelle Verifizierung enthalten ist. Mit Hilfe derartiger Methoden kann eine bestimmte Sequenz von Versuchspunkten festgelegt werden, was sich in allgemeiner Form in folgender Gleichung darstellen läßt:

$$x^{i+1} = x^i + h^i \cdot d^i$$

x^i Wert des nächsten Schrittes,
h^i Konstante,
d^i Vektor, der die Richtung der Optimierung angibt.

Tabelle 1. Beispiele von Modellbildung (Strategien)

Prozeß	Prozeßcharakteristik Problemstellung	Parameter	Strategie	Ergebnisse	Literatur
Phenolhaltige Abwässer (*P. putida*)	Toxizität des Substrates, Prozeßcharakterisierung mittels zweier Zustandsvariabler (Phenolkonzentration, Biomasse) nicht möglich	Biomasse, Phenolkonzentration, Phenolaufnahmerate	Phenolaufnahmerate als zusätzliche Zustandsvariable eingeführt	Veränderungen der physiologischen Aktivität der Kultur mit KALMAN-Filter verfolgt	[3]
Belebtschlammverfahren	Verhalten des Belebtschlammes bei Änderung der Substratkonzentration. Hysterese der Wachstumsrate (keine Übereinstimmung mit MONOD-Kinetik)	CO_2-Produktionsrate	Angenommene Parameter: Enzymmenge Enzymzerfall optimale Enzymmenge Gleichgewichtskonzentration gemäß MONOD-Kinetik	Auf Modellabwässern gute Übereinstimmung mit experimentellen Ergebnissen	[4]
Belebtschlammverfahren Pilzbiomasse (Pekilo-Protein) (*B. thuringiensis*) (Exotoxinproduktion)	MONOD-Kinetik zur Parameter- und Statuserfassung nicht geeignet	O_2-Aufnahmerate	Zusatzparameter: Altersverteilung der Biomasse	Altersverteilungsmodell bei Wachstum verifizierbar, nicht bei Belebtschlamm. KALMAN-Filter beschränkt anwendbar (1 Zustandsvariable, O_2), Kurvenanpassungsverfahren zur Parameterermittlung.	[5]
Belebtschlammverfahren	Problematik durch Variation des Zulaufs (Menge und Qualität)		Massebilanzen. Anwendung verschiedener kinetischer Gleichungen: MONOD [7] TEISSIER [8] MOSER [9] MIKESELL [10]	Spitzenanfall des Abwassers kann vorausgesagt werden	[6]
Belebtschlammverfahren	Schwer abbaubare Substanzen (Tetrapropylenbenzensulfonsäure)		Abbaukinetik unter Einbeziehung des endogenen Metabolismus	Modell gut verifizierbar	[11]

Tabelle 1: (Fortsetzung)

Prozeß	Prozeßcharakteristik Problemstellung	Parameter	Strategie	Ergebnisse	Literatur
Belebtschlamm-verfahren	Kinetik des O_2-Verbrauchs korreliert schlecht mit Substrataufnahme	OTR (Atmung) und kumulativer O_2-Verbrauch bestimmt	Transport- und Substrat-speichervorgänge im Modell implementiert	Zustandsparameter iterativ bestimmt. Modell kann Variation im O_2-Verbrauch verarbeiten	[12]
Abbau organischer Abläufe	Abbau durch MONOD-Kinetik inadequat beschrieben		CONTOIS' Wachstumsmodell. Ausbeute sinkt wegen Erhaltungsenergie	Kinetik nach CONTOIS [13] für aeroben und anaeroben Abbau geeignet	[14]
Backhefeproduktion	Optimale Führung der Fermentation bei kritischer Glukose- und Gelöst-O_2-Konzentration	Substrat- und O_2-Aufnahmerate	Exponentielle Erhöhung der Zulaufrate	Steigerung von μ von 0,22 auf 0,275 h^{-1}	[15]
Produktion von Sekundärmetaboliten (Penicillin)	Katabolitregulation bei Sekundärmetabolitproduktion		CONTOIS' Wachstumsmodell	Maximale Produktbildung ist relativ unabhängig vom Zulaufschema	[16]
Antibiotikaproduktion (Penicillin)	Modell zur Ermittlung optimaler Temperaturprofile	Temperatur	Logistisches Wachstumsgesetz	Gute Übereinstimmung mit Experiment. Optimierung gemäß Modell	[17]
Antibiotikaproduktion (Gramicidin S)	Produktbildung ist Altersfunktion, Produkt von reifen Zellen gebildet	Biomasse, Gramicidin S	Kultur: reife und unreife Zellen, Alterung = Bildung reifer Zellen	Gute Übereinstimmung mit Experiment auch Übergangszustände	[18]

Das Produkt $h^i \cdot d^i$ gibt die Größe und die Richtung der Optimierung an. Es handelt sich dabei größtenteils um ein „gezieltes oder gerichtetes Suchen". Da der Rechenaufwand ziemlich groß ist, erhielten diese Methoden, die bereits teilweise während des 2. Weltkrieges entwickelt wurden, durch den Digitalcomputer einen enormen Aufschwung.

In der Folge seien an einigen Beispielen die in der Fermentationstechnologie angewendeten Optimierungsstrategien gezeigt. NYESTE et al. [31, 32] und VERES et al. [33] minimierten (optimierten) die Verdoppelungszeit von *Hansenula anomala* mit Hilfe einer von NELDER und MEAD [34] modifizierten Simplex-Methode. Die Suche nach dem Optimum beginnt mit der Aufstellung eines Polyeders (Simplex) aus den unabhängigen Variablen. In diesen Punkten werden die zu optimierenden Parameter ermittelt. Der Punkt mit dem schlechtesten Wert wird gespiegelt, wodurch ein neuer Simplex entsteht. Dieses Verfahren wird wiederholt, bis die Optimalwerte erhalten werden. Gegenüber dem Ausgangswert konnte die Verdoppelungszeit von *H. anomala* um ca. 50% verringert werden. NELDER und MEAD [34] gehen dabei nach bestimmten Ausdehnungs- und Kontraktionsschritten vor.

Nach SWANN [35] ist die Methode bei mehr als fünf Variablen nicht mehr effizient genug.

UNDEN und HEDEN [29] verwenden in ihrer On-line-Optimierung (pH-Wert und Temperatur) des Wachstums von *Methylomonas methanolica* POWELL's Methode [36], kombiniert mit der von HOOKE und JEEVES [37]. Man verwendet einen Satz richtungsgebender Vektoren für die Suche nach neuen Bedingungen. HOOKE und JEEVES [37] kombinieren wählbare und festgelegte Suchbewegungen.

Die Optimierung war nicht erfolgreich. Das Versagen muß aber auf systembedingte Mängel wie das Fehlen von Sensoren zur frühzeitigen Erkennung von Wachstumshemmungen, zu flache response-Flächen oder die Reaktionsträgheit der Mikroorganismen zurückgeführt werden und nicht auf die angewendeten Methoden.

Methoden, die mehr auf der chemischen Prozeßtechnik basieren, wie die Optimierung des OTR nach der dynamischen K_La-Bestimmung eignen sich besser zur On-line-Analyse [30, 31]. Eine weitere Optimierungsmethode ist die 1951 von BOX und WILSON [38] entwickelte, auch Methode des steilsten Anstieges genannt. SCHRÖDER und WEIDE [39] haben mit dieser Methode die Wachstumsgeschwindigkeit von *C. guilliermondii* von 0,12 auf 0,53 h^{-1} erhöht und zwar durch Optimierung der Nährstoffgaben. REMMERS [40] wiederum fand die Methode, angewendet zur Optimierung von Wachstum und Belüftung, weniger effizient als eine einfache sequentielle Technik, bei der jeweils eine Variable geändert wird.

Durch Erarbeitung eines Temperaturprofils [41] konnte die Penicillin-Ausbeute um ca. 15% gesteigert werden. Wie die Autoren betonen, müßten zur Optimierung eines Prozesses die Profile aller Kontrollvariablen optimiert werden. Dies bedeutet, daß man vom Modell her alle Variablen erfassen müßte. Dazu wäre ein umfassenderes Wissen über die Kinetik und Transportvorgänge der einzelnen Prozesse notwendig.

2. Prozeßkontrolle in der technischen Mikrobiologie

Die Aufgabe der biotechnischen Prozeßkontrolle besteht im wesentlichen in der Datenaquisition sowie in der Steuerung und Regelung. Aus der Sicht des Elektronikers besteht zwischen der chemischen und biotechnischen Verfahrenskontrolle kein allzu großer Unterschied. Beim biologischen Prozeß besteht das Problem darin, daß er in hohem Maße interaktives Verhalten zeigt. Der Fermentationstechnologe muß mit Hilfe einer meß- und regeltechnischen Ausrüstung, die für relativ einfache Prozesse (z. B. chemische) entwickelt wurde, einen sehr komplexen Prozeß führen. Trotz der enorm fortgeschritte-

Tabelle 2. Übersicht über die in der Fermentationstechnik hauptsächlich eingesetzten Prozeßparameter

Parameter	Meßprinzip	Ableitung physiologisch relevanter Eigenschaften	Bemerkungen Literatur
Physikalische Parameter			
Temperatur	Widerstandsthermometer Thermistoren	möglich über Kalorimetrie	[45]
Volumen und Gewichtsmessungen	Dehnungsmeßstreifen (DMS) Kapazitätsmessung		[45]
Flüssigkeitsmengenmessung	Faraday. Induktionsgesetz Ultraschall		
Gasmengenmessung	thermisches Massenflußmeter		
Viskosität	Kugelfallviskosimeter (registrierend)	wäre möglich, z. Z. nicht angewendet	[46]
Dichtemessung	Biegeschwingermethode	wie oben	[47]
Kraftaufnahme	Torsion (DMS), elektrisch		
Drehzahl	Tachometer, elektrisch		Rührgeschwindigkeit wäre aussagekräftiger
Chemische Parameter			
pH-Messung	Glaselektrode	möglich (z. B. Säureproduktion)	
Ionensensitive Elektroden	Glaselektroden mit selektiven Membranen	möglich	[48]
Redoxpotential		möglich, bei gleichzeitiger Messung von pO_2 und pH	[49]
Massenspektrometrie		signifikante Aussagen über physiologischen Status, z. B. Atmung	schnelle Ansprechzeit [50]
O_2-Analysator	paramagnetisch	wie oben	langsamere Ansprechzeit als Massenspektrometer, Drift
CO_2-Analysator	infrarotspektrometrisch	wie oben	wie oben
Gelöst-O_2	Elektroden (Clark-Typ) Tubing Methode	möglich	Anwachsen von Mikroorganismen Membran, Drift

Gelöst-CO_2	pH-Elektrode mit selektiver Membran	möglich	wie oben
Autoanalyzer-Systeme		möglich	Analysendauer nachteilig (bis 20 min)
Trübungsmessung		möglich	in situ Messung problematisch, Durchflußphotometer [51]
Sensor f. flüchtige organische Verbindungen	Widerstandsänderung einer SnO—ZnO—SiO-Perle bei Kontakt mit leicht oxydierbaren Substanzen	möglich (z. B. Steuerung einer Backhefegärung)	Nachteil: Querempfindlichkeit [52]
Biologische Parameter			
Fluorometrie	NADH durch Fluoreszenz bestimmt	bedingt möglich	keine eindeutige Korrelation NADH-Biomasse [53]
ATP	Biolumiszenz-Methode [54], [55] (1964)	sehr bedingt möglich	keine eindeutige Korrelation zwischen Atmungstätigkeit und ATP-Gehalt [56] [57]
Enzymelektroden	Transduktorgebundene Enzyme	gut möglich, hohe Spezifität	Standzeit der Enzyme problematisch, Messung im by-pass [58]
Mikrobielle Elektroden	Immobilisierte ganze Zellen kombiniert mit O_2-Elektrode, mit Brennstoffzelle und mit Membranelektroden (pH, pCO_2)	gut möglich	ähnliche Nachteile wie Enzymelektroden [59—61]
Enzym-Thermistoren	Messung der bei Enzymreaktionen freiwerdenden Wärme	universell einsetzbar, spezifisch	[62]
Enzym-Transistoren	NAD/NADH gekoppelte Reaktionen	möglich	auf Dehydrogenasereaktion beschränkt [62]
Affinitätssensoren	FITC-Dextran d. Liganden freigesetzt und detektiert	bedingt möglich	Spezifität des Bindungsagens ist entscheidend [63]
Abgeleitete Prozeßparameter			
Atmung		signifikante Aussagen über physiologischen Status	schnelle Anwortfunktion des biologischen Systems
Dynamische Kalorimetrie		wie oben, universell einsetzbar	mangelnde Spezifität

nen biochemischen Erkenntnisse können die Vorgänge auf metabolischer Ebene oftmals nicht quantitativ und in ihrer Komplexheit erfaßt werden. Meist fehlen auch die geeigneten Sensoren.

2.1. *Probenahme und Probenaufbereitung*

Die Implementierung der Sensoren hat meist in situ und unter aseptischen Verhältnissen zu geschehen. Dadurch mußten erst sterilisierbare Sensoren und/oder aufwendige Konstruktionen zur aseptischen Probenahme entworfen werden.
MARTEN [42] beschreibt ein 3-Ventil-System zur aseptischen Probenahme. Das System hat noch nicht den zur On-line Analyse notwendigen Automatisierungsgrad erreicht. In letzter Zeit ist ein deutlicher Trend zur Messung im by-pass zu vermerken, was den Einsatz nicht sterilisierbarer Meßeinrichtungen möglich macht. Besonders gedacht ist hier an enzymgebundene Transduktoren. Ähnlich der von MARTEN [42] beschriebenen Dialyse bzw. Filtration der Probe wurde von NEESTAS und WANG [43] die „Filtrationsprobe" zur Bestimmung der Biomasse beschrieben. Gedanklich fallen auch die im Englischen als „tubing"-Methoden beschriebenen Verfahren in dieses Kapitel. YANO et al. [44] beschreiben dieses System zur Methanolbestimmung in fed-batch-Kulturen.

2.2. *Prozeßparameter*

Der Entwurf einer Kontrollstrategie hat primär von der Identifizierung relevanter Prozeßparameter auszugehen. Die Messung selbst macht sich meist einfache physikalische oder chemische Effekte zunutze, die in ein elektrisches Ausgangssignal umgewandelt werden, was die Grundlage für eine weitgehende Automatisierung schafft. Diese meist einfachen Messungen von z. B. Temperatur, Druck, Viskosität und Stofftransport charakterisieren einen mikrobiologischen Prozeß meist in nur sehr geringem Maße. Es fehlt die On-line-Erfassung biologischer Schlüsselparameter. Dies ist nur über Umwegparameter (gateway-Sensoren) möglich wie z. B. Wärmetönung, O_2-Verbrauch bzw. CO_2-Bildung.
Die in der Folge kurz dargestellten Parameter sind aus Gründen der besseren Übersichtlichkeit in Tabelle 2 zusammengefaßt. CONSTANTINIDES et al. [17] teilen Prozeßparameter in Zustandsvariable und Kontrollparameter ein. Eine Zustandsvariable beschreibt den Zustand des Systems. Ein direkter Zugriff auf sie ist nicht möglich. Zu den Kontrollvariablen hat der Experimentator direkt Zugriff.

2.2.1. Erfassung biologischer Parameter

Ergänzend zu den in Tabelle 2 angeführten Parametern werden in der Folge noch jene mit physiologisch relevanter Antwortfunktion behandelt.

2.2.1.1. Fluorometrie

Der Vektor ist das durch Fluoreszenz bestimmte NADH; er wurde verschiedentlich zur Bestimmung der Biomasse vorgeschlagen (u. a. ZABRISKIE [64]). BEYELER et al. [53] stellten fest, daß die Biomasse nur während der exponentiellen Phase mit der Fluoreszenz korreliert.

2.2.1.2. ATP-Bestimmung

Auf Grund der Schlüsselstellung des ATP versucht man, es intensiv zur Charakterisierung der Stoffwechselaktivität, insbesondere der des Belebtschlammes, heranzuziehen (ca. 340 Publikationen [65]). Der ATP-Gehalt wird nach der Biolumineszenz-Methode bestimmt [54, 55]. ATP ist bekanntlich einer ständigen Umsetzung unterworfen. In einer *E. coli* Zelle werden pro Sekunde $2{,}5 \cdot 10^6$ Moleküle ATP zu ADP und Phosphat abgebaut [56]. THOMANETZ et al. [56] fanden ATP-Gehalte von 0,2—2 mg ATP/g Belebtschlamm-TS. Die analysierte ATP-Menge resultiert letztlich aus einem Pool. THOMANETZ et al. [57] fanden auch keine Korrelation zwischen ATP-Gehalt und Atmungsaktivität.

2.2.1.3. Enzymelektroden

Allgemein bezeichnet man derartige Systeme als „Transduktorgebundene Enzyme" [58]. Immobilisierte Enzyme, auf einer semipermeablen Membran aufgebracht, bewirken an dem zu bestimmenden Substrat eine physikalische oder chemische Veränderung, die vom Transduktor in ein elektrisches Signal umgewandelt wird.
Die Spezifität hängt von der des Sensors und der Enzyme ab. MOR [59] berichtet von ca. 40 verschiedenen Anwendungen, wovon einige bereits marktfähig gemacht wurden. SUZUKI und KARUBE [60] beschreiben u. a. Enzymelektroden für Saccharose, Cholesterin, Phospholipide und Harnsäure. DANIELSSON und MOSBACH [62] beschreiben eine sterilisierbare Penicillin-Elektrode.
Nachteilig sind der hohe Preis und die mangelnde Stabilität, was schließlich zur Entwicklung der mikrobiellen Elektroden führte.

2.2.1.4. Mikrobielle Elektroden

Das Prinzip ist den Enzymelektroden analog. Im wesentlichen haben sich drei Kategorien herauskristallisiert [59].

1. Immobilisierte ganze Zellen kombiniert mit einer O_2-Elektrode
2. Immobilisierte ganze Zellen oder Zellsuspensionen kombiniert mit einer Brennstoffzelle
3. Immobilisierte ganze Zellen kombiniert mit Membran-Elektroden (pH, pCO_2)

Eine bekannte Anwendung ist die BSB-Elektrode [66, 67]. Nachteilig sind auch hier der Preis und die doch kurzen Standzeiten.
KOBOS und PYON [68] entwickelten eine mikrobielle Elektrode zur Bestimmung von Nitriltriessigsäure, die in 4 Schritten durch Decarboxylierung und Desaminierung zu Glycerat abgebaut wird. Der freigesetzte Ammoniak wird mit einer Ammoniumelektrode bestimmt.

2.2.1.5. Enzym-Thermistoren

Bei dieser Methode, entwickelt von DANIELSSON et al. [69], wird die bei enzymatischen Reaktionen frei werdende Wärme gemessen. Die Spezifität der Messung ist durch das verwendete Enzymsystem gegeben. Der lineare Meßbereich ist sehr weit, z. B. 0,01-250 m Mol bei Penicillinase und Urease [62].

2.2.1.6. Enzym-Transistoren

Diese Methode stellt die neueste Entwicklung unter den Biosensoren dar [62]. Der von einer Hydrogenase aus NADH spezifisch freigesetzte Wasserstoff wird mittels eines Halbleiterbausteines bestimmt.

2.2.1.7. Affinitäts-Sensoren bestimmter Metaboliten [63]

Die Prinzipien zur Entwicklung von Affinitätssensoren sind ähnlich jenen der Radio-Immun-Assay-Technik. Die Metaboliten werden aber mit optischen Methoden bestimmt (Fluoreszenzmessung, FITC-Dextran).

2.2.2. Abgeleitete Prozeßparameter

In dieser Gruppe werden im wesentlichen zwei Parameter zusammengefaßt, die Aussagen über den metabolischen Status, hauptsächlich von aeroben Mikroorganismen, zulassen. Es sind dies die Atmung und die Kalorimetrie. Aus der Atmung bzw. aus der O_2-Aufnahmerate und CO_2-Produktionsrate kann nach der Methode der Komponentenbilanzierung [70, 71] auf die gebildete Biomasse geschlossen werden.

2.2.2.1. Atmung

Die Atmung wird mittlerweile sehr häufig zur Charakterisierung der physiologischen Aktivität der Biomasse eingesetzt. Dies ist auf die hohe physiologische Relevanz des Parameters, auf die Einfachheit und Geschwindigkeit der Messung zurückzuführen.
Die Hauptrolle des Sauerstoffs ist die des endterminalen Elektronenakzeptors. Bei Prokaryonten wird nur ein minimaler Anteil des molekularen Sauerstoffs in die Biomasse eingebaut, der Großteil stammt aus H_2O und CO_2 [72]. Ähnlich ist es bei Eukaryonten; bei Hefen wird molekularer Sauerstoff zur Fettsäuresynthese benötigt. Nach HARRISON [72] wäre es sinnvoller, beispielsweise statt der kritischen Gelöst-O_2-Konzentration die jeweilige Wirkungsebene des Sauerstoffs zu definieren.
Mittels der Atmung können Veränderungen im Metabolismus On-line festgestellt werden, so daß zahlreiche Konzepte einer Prozeßsteuerung praktisch verwirklicht wurden [73 bis 80]. Anfänglich wurde die gebildete Biomasse aus dem O_2-Verbrauch und dem spezifischen O_2-Bedarf errechnet [74, 81]. Da sich letzterer in Abhängigkeit von der Wachstumsrate ändert, wurde die Methode der Komponentenbilanzierung eingeführt [70, 71]. Im Labor des Autors konnte deutlich gezeigt werden [82, 83], daß der spezifische O_2-Bedarf in Abhängigkeit von der Biomasseproduktivität ein Minimum zeigt (Abb. 1). Gemäß dem in Abb. 1 dargestellten Verhalten der Mikroorganismen kann man versuchen, die Kurve abschnittsweise zu unterteilen und dafür die jeweiligen Korrelationen

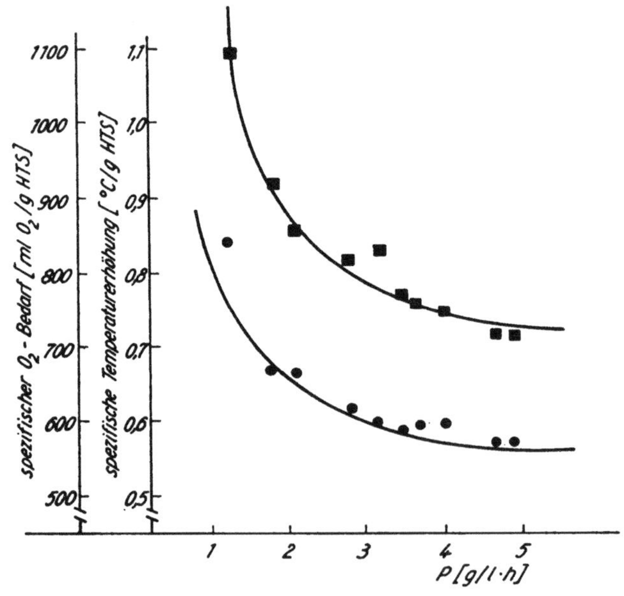

Abb. 1. Spezifischer O_2-Bedarf (■) und spezifische Temperaturerhöhung (●) von *C. intermedia* gegen Biomasseproduktivität

aufzustellen oder mit Hilfe eines Approximationspolynoms den spezifischen O_2-Bedarf (SOB) zu berechnen, das z. B. für *Candida intermedia* folgendermaßen lautet

$$SOB = 2{,}67 - 1{,}94 \cdot P + 0{,}84 \cdot P^2 - 0{,}168 \cdot P^3 + 0{,}012 \cdot P^4$$

P = Biomasseproduktivität

Weniger empirisch ist die Komponentenbilanzierung [70, 71]. Diese Methode hat sich inzwischen in der Fermentationstechnik fest etabliert. Sie wurde u. a. von WANG et al. [75] zur Steuerung einer Zulauffermentation von Backhefe eingesetzt. BRAVARD et al. [84] beschreiben ihren Einsatz zur Kontrolle kontinuierlicher Kulturen von Hefe auf Melasse. Durch Änderungen der molekularen Zusammensetzung der Biomasse in Abhängigkeit von der Wachstumsrate kann es zu Abweichungen kommen. Bekanntlich steigt der RNS-Gehalt mit der Wachstumsrate linear an; bei *C. intermedia*, kultiviert auf Molke [85], zeigten sich bei bestimmten Verdünnungsraten Abweichungen von der Linearität.

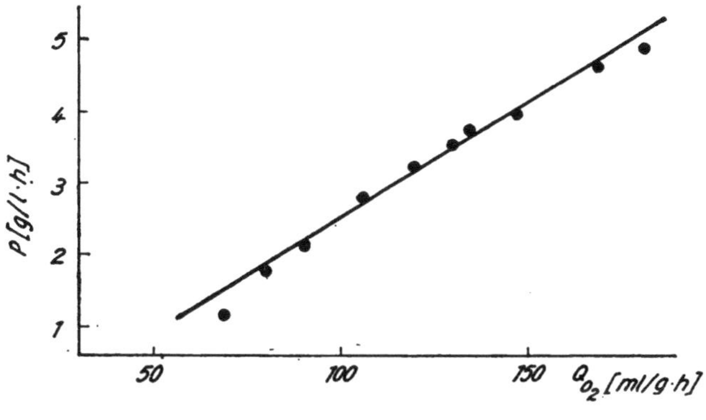

Abb. 2. Biomasseproduktivität (P) gegen spezifische Atmungsrate (Q_{O_2})

In eigenen Versuchen konnte im Labormaßstab unter konstant gehaltenen Fermentationsbedingungen (aerob) zwischen spezifischer O_2-Aufnahmerate (Q_{O_2}) und Biomasseproduktivität eine vollkommen positive Korrelation ($r = 0{,}995$) [83] festgestellt werden. Für den Fall, daß der RQ-Wert von 1 abweicht, ist es notwendig, die aus der Abluftanalyse via Computer für den O_2- und CO_2-Gehalt erhaltenen Werte nach der von FIECHTER und von MEYENBURG [86] erarbeiteten Formel zu korrigieren.

Verhalten des O_2-Bedarfes bei pulsartiger Substratzufuhr

Kontinuierliche Kulturen von *C. intermedia* wurden mit konzentriertem Substrat gepulst und das Verhalten der Atmung beobachtet [20, 21]. Die Hefe nimmt die gepulste Saccharose sehr rasch auf (anfängliche Aufnahmeraten von 100 mg · g^{-1}HTS · min^{-1}). Die Aufnahmekinetik zeigt Sättigungsverhalten. Nach 4 min. war die spezifische Aufnahmerate auf 20 mg · g^{-1}HTS · min^{-1} gesunken. Ähnliches Verhalten zeigt sich, wenn innerhalb von 5—10 min. ein zweites Mal gepulst wird (Tabelle 3).
Saccharose wird direkt aufgenommen, was im Gegensatz zu den von SOLS und de la FUENTE [87] aufgestellten Modellen steht, bei denen Saccharose vor der Aufnahme extrazellulär hydrolysiert wird. Ein prinzipiell ähnliches Verhalten von *Klebsiella aerogenes* nach Substratpulsen wurde von LEEGWATER et al. [88] festgestellt. Es ist daher zu beachten, die Substratzufuhr möglichst gleichmäßig zu gestalten. Dies hat besonders bei Verfahren mit stoßartigem Substratanfall (z. B. Abwassertechnologie) Bedeutung.

Tabelle 3. Wachstumskinetik unter Gleichgewichtsbedingungen und nach Pulsen

	Gleichgewichtsbedingungen	Puls
Saccharose-Aufnahmerate (mg/gHTS · min)	10,9 (bei max. Biomasse-produktivität)	13,5 (durchschnittl. Aufnahme 10 min nach Puls)
max. Saccharose-Aufnahmerate (mg/gHTS)	—	100 (1 min nach Puls)
Gesamt O_2-Verbrauch während des Pulses (mg/l, 30 min)	948 (spez. O_2-Bedarf 0,9 g/gHTS)	1335
Hefetrockensubstanz (HTS g/l)	15,6	16,8 (30 min nach Puls)
Ausbeute (g Biomasse /g Sacch. verbraucht)	51%	52%
max. Biomasseproduktion	5	8,5 (berechnet aus Sacch. Aufnahme, Ausbeute 51%)

Bei der Ableitung des Wachstums von Mikroorganismen aus dem O_2-Verbrauch muß korrekterweise auch der endogene O_2-Verbrauch einbezogen werden, der in den oben dargestellten Korrelationen bereits berücksichtigt ist.
Der Anteil an endogener Atmung kann nach einer von Herbert (89) aufgestellten Gleichung $Q_{O_2} = \alpha \cdot \mu + \beta$ berechnet werden, wobei β den Anteil für den Erhaltungsstoffwechsel darstellt, der bei $\mu = D = 0$ berechnet werden kann.

2.2.2.2. Kalorimetrie

Die bei metabolischen Reaktionen entstehende Wärmemenge stellt einen sehr universellen Parameter zur Charakterisierung mikrobiologischer Vorgänge dar. Stark abstrahiert kann man die im aeroben Zellstoffwechsel gebildete Wärmemenge auf die Substratkettenphosphorylierung, die Atmungskettenphosphorylierung und die bei Biosynthesen freiwerdende Wärmemenge zurückführen. Der Hauptanteil resultiert aus den letzten beiden Vorgängen. Forrest und Walker [90] errechneten, daß 71% der in Form von ATP für Synthesen verfügbaren Energie zur Proteinsynthese verwendet werden, wovon 75% als Wärme abgegeben werden. Die Quantifizierung der Wärmetönung müßte in den einzelnen Materialbilanzen stärker berücksichtigt werden. Nachteilig ist die mangelnde Spezifität der Messung. Die geringe Verbreitung der Methode ist vielleicht auch auf gewisse Umständlichkeiten bei der Messung zurückzuführen. Die auf dem Markt befindlichen Durchflußkalorimeter erlauben nur Messungen im by-pass. Bereits 1968 beschreiben Cooney et al. [91] eine in situ Messung nach der Methode der dynamischen Kalorimetrie. Ähnlich dem von Danielsson et al. [69] beschriebenen Enzym-Thermistor ist der Reaktor (Fermentor) die Meßzelle. Die Temperaturerhöhung der Maische wird gemessen. Diese hängt von der durch die metabolischen Reaktionen entstehenden Wärmemenge und auch von rein physikalischen Vorgängen, wie von der durch den Rührer eingetragenen Wärmemenge, von der Abstrahlung an die Umgebung, vom Wärmeverlust durch Verdampfen von Wasser in die Abluft und von der Umgebungstemperatur ab.
Die Einflüsse externer Faktoren werden in einem Blindwert berücksichtigt.

$$Q_F = K \cdot T \qquad K = \text{Wärmekapazität des Systems (konstant)}$$

Die Wärmetönung im Fermentor korreliert direkt mit der Temperaturerhöhung. Nach umfangreichen Untersuchungen [83] wurde festgestellt, daß eine Meßzeit von 7 min. eine

ausreichend signifikante Temperaturerhöhung ergibt. Der Korrelationskoeffizient zwischen der Temperaturerhöhung (T) und der Biomasseproduktivität (P) lag bei $r = 0{,}993$. Zwischen O_2-Verbrauch und Temperaturanstieg wurde eine ebenso enge Korrelation gefunden, was sich auch in der Parallelität folgender Kurven zeigt: spezifische Temperaturerhöhung (°C/g HTS) gegen Biomasseproduktivität und spezifischer O_2-Bedarf gegen Biomasseproduktivität (Abb. 1). Aus den obigen Beziehungen läßt sich eine multiple Korrelation zwischen Biomasseproduktivität, Temperaturerhöhung und dem O_2-Gehalt der Abluft ableiten [83].
Da die Hauptmenge der produzierten Wärmemenge aus der Atmungskettenphosphorylierung und aus Synthesen stammt, wird durch eine energetische Entkoppelung (ATP-Biomasseproduktion) das Verhältnis Wärmetönung/O_2-Verbrauch verändert. Abweichungen von den oben aufgestellten Beziehungen stellen ein einfaches Mittel dar, um Verschiebungen in der Energetik des Metabolismus feststellen zu können.
COONEY et al. [91] verwendeten die dynamische Kalorimetrie zur Charakterisierung des Wachstums von *E. coli, B. subtilis, C. intermedia* und *A. niger*. MOU und COONEY [92] setzten obige Methode in der Prozeßkontrolle der Novobiocin- und Cellulaseproduktion ein. WANG et al. [93] setzten sie bei Backhefefermentationen, insbesondere zur Unterscheidung zwischen Wachstum und EtOH-Bildung ein. FORTIER et al. [94] versuchten mit gutem Erfolg auf Modellabwässern die physiologischen Vorgänge in Belebtschlammanlagen zu erfassen. PLASSE et al. [95] verfolgten den Zelluloseabbau von Mischkulturen mittels Durchflußkalorimeter in Abhängigkeit von der Vorbehandlung der Zellulose.

3. Rechnerunterstützte Prozeßkontrolle

Die mikrobiologische Prozeßkontrolle bezog ihre anfänglichen Erfahrungen aus der chemischen Verfahrenstechnik.
Die an einer einfachen Fermentation anzutreffende Meß- und Regeltechnik ist meist auf analoge Regelkreise beschränkt. Im Minimum ist eine Temperatur-, pH-, Drehzahl-, Volumen- und Niveau-Regelung vorhanden. Die anfallenden Signale werden in einschleifigen Regelkreisen dezentral, d. h. jedes Gerät für sich, verarbeitet.
Die Ansprüche des Mikrobiologen an die Prozeßkontrolle sind in den letzten Jahren durch den ständigen Zuwachs an Erkenntnissen über Vorgänge auf molekularbiologischer Ebene und deren effektive Manipulierbarkeit durch gentechnische Eingriffe deutlich gestiegen. Diese können meist nur durch eine verfeinerte Prozeßkontrolle physiologisch ausgespielt werden. Gleichzeitig ermöglichte der Aufschwung der Mikroelektronik wesentliche Elemente der digitalen Steuerungstechnik kostengünstig im Laborpilot- und Industriebereich einzusetzen. Die speicherprogrammierbare und auch die auf Mikrocomputer basierende Steuerung ist ab ca. 30 Aus- und Eingängen kostengünstiger als die konventionelle Relaistechnik [96]. Neben den allseits bekannten Vorteilen der Mikroprozessortechnik bietet die computergekoppelte Prozeßkontrolle die Möglichkeit der Selbstkalibrierung. SWARTZ und COONEY [97] führten täglich softwaremäßig eine Kalibrierung der Gasanalysatoren durch. Die Entwicklung der rechnergekoppelten Fermentationstechnik kann an den abgehaltenen spezifischen Symposien abgelesen werden. Ohne auf Einzelheiten einzugehen, seien hier nur die jeweiligen Schwerpunkte herausgegriffen. Das erste 1973 in DIJON [98] zeigte, daß der Computer bereits einen fixen Platz in der Fermentationstechnologie besitzt. Die optimistischen Voraussagen hinsichtlich seiner Weiterverbreitung erfüllten sich jedoch nur in geringem Maße. Die Zuwachsraten, mit denen in der chemischen bzw. petrochemischen Industrie Prozeßrechner installiert wurden — COONEY [99] berichtet von exponentiellen Zuwächsen von 0,043 pro Jahr — konnte die Fermentationsindustrie aus bereits erwähnten Gründen bei weitem nicht aufweisen.

Tabelle 4. Übersicht über computerkontrollierte Fermentationsprozesse

Prozeß	Organismus	Kontrollstrategie	Zielvorstellungen	Bemerkungen — Literatur
Backhefe (Labormaßst.)	*S. cerevisiae*	Fed-batch, Substratzulauf über RQ gesteuert	Grundlagen für Steuerung erarbeitet	[103]
Backhefe (Labormaßst.)	*S. cerevisiae*	Fed-batch, Substratzulauf über EtOH-Konzentration (Teflon Tubing)	Zulaufsteuerung PID und EIN/AUS Regelung, $\mu = 0{,}3$	[104]
Backhefe (Labormaßst.)	*S. cerevisiae*	Ermittlung des Quasi Steady State einer fed-batch Kultur	Substratzulauf über Offline-Analysen gesteuert	[105]
Backhefe (Labormaßst.)	*S. cerevisiae*	Zulaufrate über EtOH-Sensor-feedback Steuerung	Glukosekonzentration (GK) > 130 mg EtOH-Bildung, bei GK = 70 mg Glukose und EtOH gleichzeitig verwertet	[78]
Backhefe (Labormaßst.)	*S. cerevisiae*	Kontinuierl. Kultur, Zulauf RQ gesteuert	Mangel: Glukoseelektrode	[106, 23]
Backhefe (Labormaßst.)	*S. cerevisiae*	Komponentenbilanzierung (Abgasanalyse)	Komponentenbilanzierung ist geeignet, $\mu_m = 0{,}25$	[77]
Backhefe (Labormaßst.)	*S. cerevisiae*	Fed-batch-DDC-Zulauf über EtOH und RQ gesteuert	Feed back Kontrolle, höhere Zellausbeute durch EtOH Kontrolle	[107]
Biomasse (SCP) Backhefe	*H. polymorpha* *S. cerevisiae*	Komponentenbilanzierung (Gelöst-O_2, O_2 im Abgas, Alkali und Substratzulauf)	Komponentenbilanzierung zur Bestimmung der Biomasse	[97] [76]
Biomasse (Pekilo Protein), ind. Maßstab	*Paecilomyces varioti*	Kontinuierl. Kultur auf Sulfitablauge. Parameter: Myzelwachstum, OTR, Altersverteilung, Proteingehalt, Substrat- und Nährstoffgaben	Erzielte Ersparnis: Energie: 15—20% Phosphorsäure: 25—30% Ammoniak: 5—10% Biomasseproduktivität: 10—15% Steigerung	[108] [109]
Antibiotika		Zulaufsteuerung: RQ und Q_{CO_2} mit Massenspektrometer, hierarchische Computerkontrolle	Abgrenzung einzelner Wachstumsphasen möglich	[50]

Backhefe Gluconsäure	*S. cerevisiae* *Ac. suboxydans*	Melassequalität Backhefeproduktion	Optimierungen (pH-Temperatur), Produktivität der Gluconsäureproduktion um 58% gesteigert	[33] [31, 32]
Belebtschlammverfahren		Parameter: Kapazitätsmessung (Biomasse), Abgasanalyse (Massenspektrometer), SPC und DDC Konzepte	Systembeschreibung	[110]
Belebtschlammverfahren		Substratabbau, Zuluft über Abgasanalyse kontrolliert	Gleichmäßige Ablaufqualität (BSB 20 mg/l)	[111]
Enzymproduktion	*Asp. niger*	Unterscheidung zwischen Wachstum und Produktbildungsphase	70% Steigerung der Enzymproduktion	[101]

Am nächsten Symposium in Philadelphia 1978 wurden einerseits neue Sensoren (z. B. Enzymelektroden, Fluoreszenzmessungen), andererseits Methoden wie Komponentenbilanzierung und Modelle zur Optimierung vorgestellt. In Pilotanlagen war der Computer bereits eine fix integrierte Komponente. Sein Einsatz war auch ökonomisch vertretbar. Das Symposium in Manchester 1981 zeigte, daß von der Hardware her eine Fülle von reinen Prozeßrechnern und Minicomputern angeboten wird und daß auch auf Seiten der Sensoren (z. B. Enzymelektroden, -thermistoren und -transistoren) Fortschritte erzielt wurden. Darüberhinaus sind derzeit bereits voll Computer-gesteuerte Industrieanlagen in Betrieb [50, 100, 101, 102]. Eine repräsentative Übersicht über die in den letzten Jahren im Bereich der Fermentationstechnik eingesetzten Prozeßrechner gibt die Tab. 4.

3.1. Prinzipieller Aufbau der rechnergekoppelten Fermentationsanlage

Den schematischen Aufbau der an den Fermentor gekoppelten Prozeßrechenanlage zeigt Abb. 3.
Vor ca. drei Jahren wurde im Labor des Autors der Mikrocomputer Commodore 3032, ein 8 bit Rechner, angeschafft. Er kann in BASIC programmiert werden und besitzt eine (IEEE 488) Schnittstelle.

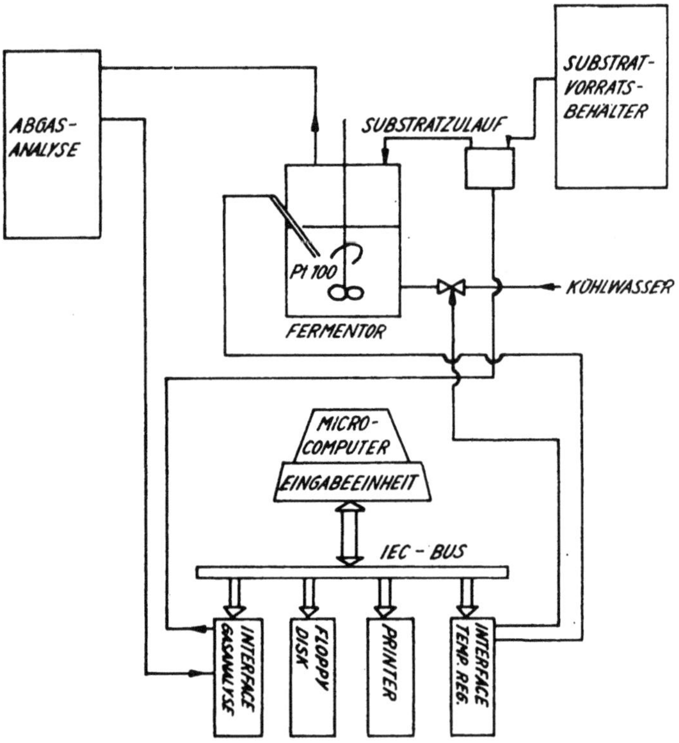

Abb. 3. Prinzipdarstellung Fermentor — Prozeßrechenanlage

3.1.1. Rechnerperipherie

Zur Daten- und Programmspeicherung wird eine Floppy Disk (Commodore 4040) mit zwei Laufwerken verwendet. Weiter kann zur Protokollierung von Meßwerten auch ein Matrixdrucker (Commodore 3022) eingesetzt werden. Beide Geräte sind via IEC-Bus

mit dem Rechner verbunden. Daran sind noch zwei Interfaces gekoppelt, zum einen (PCI 2ai 4ao) mit 2 Analogeingängen (0-20 mA) und 4 Analogausgängen (0-30 V Gleichstrom), das zur Verarbeitung der in der Abgasanalyse anfallenden Daten (% O_2 und CO_2) und zur Steuerung diverser Pumpen (u. a. Substratzulauf und -ablauf) verwendet wird und zum anderen (PCI 1ai 4ao) mit einem Analogeingang für ein Widerstandsthermometer (Pt 100) und vier Analogausgängen (0-30 V Gleichstrom) und 4 Relais, das zur Ansteuerung von Magnetventilen verwendet wird.
Beide Interfaces sind voll programmierbar. Die Auflösung der Analogeingänge beträgt auf Grund der 8 bit Adressierung 4×10^{-3}. Im konkreten Fall liegt sie bei ersterem Interface bei 10^{-2} mA und bei letzterem bei 10^{-2} °C. Die Auflösung der Analogausgänge beträgt 256 Schritte (2^8 bits).

3.2. *Aufgaben und Anforderungen an den Prozeßrechner*

Der Fermentationstechnologe stellt im wesentlichen an den Prozeßrechner folgende Anforderungen: Meßwerterfassung, Meßwertverarbeitung, Datenanalyse, Systemüberwachung, Protokollierung und Datenspeicherung.
Davon seien die folgenden Punkte herausgegriffen:

3.2.1. Meßwertverarbeitung

Ein nun in digitaler Form zur Verfügung stehender Meßwert muß zur weiteren Nutzung in die entsprechenden physikalischen Einheiten umgerechnet werden.
Besteht keine lineare Abhängigkeit, so wird man versuchen, andere mathematische Funktionen aufzufinden (z. B. Polynome, logarithmische- oder Winkelfunktionen).
Neben systematischen Fehlern können auch zufällige auftreten, die eliminiert werden müssen. Diesen Vorgang nennt man digitale Filterung oder Glättung. Neben der Mittelwertbildung, die aber nur bei konstanter Meßzeit erlaubt ist, wurde von BERNARD et al. [112] eine relativ einfache Methode entwickelt. Aus einer bestimmten Anzahl von Meßwerten wird jeweils der Mittelwert gebildet, wobei Extremwerte verworfen werden. Eine weitere noch relativ einfache Methode ist die Anwendung von Spline-Funktionen [113]. Aus einer Anzahl von Punkten wird ein Interpolationspolynom gebildet, aus dem wiederum ein Polygonzug mit minimaler Schwankung aber mit Unstetigkeitsstellen gebildet werden kann (Spline). Aufwendigere Methoden werden von JEFFRIS et al. [114] beschrieben. Mittels des Filters wird die Zellmasse ($x(t)$) und die Wachstumsgeschwindigkeit (dx/dt) aus einer Reihe von Messungen (Zellmasse $x_1, x_2 \ldots x_n$) wiederholt aufgenommen und zur Zeit $t_1, t_2 \ldots t_n$ nach der Methode der kleinsten Quadrate bestimmt.
SVRCEK et al. [115] verwenden die KALMAN-Filtertechnik zur Modellierung kontinuierlicher Kulturen. Die Problematik besteht darin, daß meist nur ein Parameter (Gelöst-O_2, OTR) zur Abschätzung der übrigen zur Verfügung steht. Durch modellbedingte Ungenauigkeiten können Störsignale nicht mehr erkannt und eliminiert werden [114].

3.2.3. Systemüberwachung

Verglichen mit konventionellen Regelungen kann der Prozeßrechner Abweichungen frühzeitig erkennen, (z. B. Grenzwerterkennung). Bei dieser Art der Fehleranalyse treten im Prozeß bereits Abweichungen vom Sollzustand auf. Um das zu verhindern, schlägt JEFFRIS [116] eine in der Flugtechnik und Elektronik angewandte Fehleranalyse vor. Bei Veränderung eines Parameters werden meist auch andere tangiert und verändert. Die Fehleranalyse wird nun aus Kombinationen bestimmter Daten und Parameter (Fehlerlexikon) vorgenommen. Unabhängig vom Prozeß wird sie ständig durchlaufen und kann somit Abweichungen vom Sollzustand kurz nach deren Auftreten bereits feststellen. Von Blachère [98] wurde eine derartige Kontrolle in einer Pilotanlage (INRA-DIJON) implementiert. Unter die systemüberwachenden Aufgaben fällt auch die periodische Kalibrierung.

3.3. Software

Mittels der Software entsteht schließlich erst die spezifische Problemlösung. Bei der Software-Entwicklung kann heute mit Hardware-Funktionen viel großzügiger umgegangen werden. Der Software kommt somit immer größere Bedeutung zu.

Die Bestrebungen, die Programmierung in genormte Bahnen zu lenken, finden in der Etablierung des Software-Engineering ihren Ausdruck. Selva und Zerbatto [117] konnten durch exakte Planung die Testphase der von ihnen entwickelten Software entscheidend verkürzen. Gegenüber der traditionellen Aufwandsverteilung (33% Planung, 17% Kodierung und 50% Tests) erreichten sie durch Intensivierung der Planung folgendes Bild: 50% für Planung, 20% für Kodierung und 30% für Tests. Die verlängerte Planungsphase hat den Vorteil, daß versteckte Fehler vermieden werden und daß alle Mitarbeiter über das Programm informiert sind, was die Wartung erleichtert. Den Wünschen nach einer strukturierten, modular aufgebauten Software stehen die teils unklaren gesetzlichen Maßnahmen zum Schutz von Softwareentwicklungen entgegen.

4. Praktische Realisierungen Mikrocomputer-kontrollierter Fermentationen

In den folgenden Abschnitten soll an einigen Beispielen die praktische Realisierung einiger Kontrollkonzepte mittels eines Personalcomputers gezeigt werden.

Dabei soll bewußt auf die Einfachheit und die daraus resultierende Preiswürdigkeit der Prozeßrechenanlage hingewiesen werden. Die Gesamtkosten für den Rechner und die Interfaces liegen bei ca. 24000 DM. Somit ist es auch möglich, im Universitätsbereich derartige Entwicklungsarbeiten zu betreiben und meßtechnisch ein Instrumentarium mit höherer physiologischer Relevanz und Antwortfunktion zu aktivieren, als dies in konventionell geführten Fermentationen der Fall ist.

4.1. Identifikation und Wichtung der Parameter Atmung und dynamische Kalorimetrie

Das Verhalten der beiden Parameter, wie sie in Kapitel 2.2.2. beschrieben sind, wurde in kontinuierlicher Kultur, teils auch mittels Pulsexperimenten, ermittelt. Die Ansprechzeit ist ein wichtiger Faktor für einen Prozeßparameter. Die Hefe *Candida intermedia* wurde auf Nährmedium I mit 5,8 g Saccharose bei 2,5 l Kulturvolumen gepulst.

Nährmedium I (Angaben in g/l)		Spurenelementlösung (Angaben/l)	
Saccharose	30	$CuSO_4 \cdot 5H_2O$	300 mg
$(NH_4)_2SO_4$	4,4	$MnSO_4 \cdot H_2O$	800 mg
$(NH_4)_2HPO_4$	1,92	$NaMoO_4 \cdot 2H_2O$	400 mg
KCl	0,87	$ZnSO_4 \cdot 7H_2O$	3,0 mg
$MgSO_4 \cdot 7H_2O$	0,45	$FeCl_3 \cdot 6H_2O$	4,0 mg
$CaCl_2 \cdot 2H_2O$	0,42		
Spurenelementlösung	1 ml		

Tab. 5 zeigt einen Vergleich des Rest-O_2-Gehaltes der Abluft und der Temperaturerhöhung unter Gleichgewichtsbedingungen mit denen während eines Pulses.
Das wesentliche dieser Gegenüberstellung ist die Tatsache, daß beide Parameter auf Störungen unmittelbar, und zwar schon nach 1 min., reagieren (vgl. jeweils Werte und Spalten 2 und 4 mit jenen von 1 und 3). Mittels beider Parameter können daher Veränderungen im Prozeß schnell erkannt werden.

Tabelle 5. Verhalten von *C. intermedia* nach Substratpulsen

Zeit (min)	Abnahme des O_2-Gehaltes in der Abluft (Absolut %)		Temperaturerhöhung (dynam. Kalorimetrie, Bruttowerte)	
	Gleichgew. Bedingungen	Puls-experiment	Gleichgew. Bedingungen	Pulsexp.
0	0	0	0	0
1	0,2	0,3	0,18	0,23
2	0,5	0,9	0,35	0,44
3	0,7	1,2	0,52	0,67
4	—	1,5	0,71	0,88
5	—	1,7	0,90	1,13
6	—	1,6	1,08	1,34
7	—	1,5	1,21	1,57

4.2. Entwicklung von Kontrollalgorithmen

Gemäß den in Kapitel 2.2.2. beschriebenen Befunden herrscht unter definierten Kultivierungsbedingungen eine enge Korrelation zwischen dem O_2-Gehalt der Abluft, der produzierten Wärmemenge und der Biomasseproduktivität. Verfahrenstechnisch eröffnet dies die Möglichkeit, den Substratzulauf in Abhängigkeit vom jeweiligen Wachstumsstadium bzw. der jeweiligen physiologischen Aktivität der Kultur zu steuern. Die Steuerung des Substratzulaufes wird vorrangig bei kontinuierlichen und bei fed-batch Kulturen von Interesse sein, beispielsweise bei folgenden Problemstellungen:

— Substratzulauf in Abhängigkeit von der jeweiligen Wachstumsgeschwindigkeit
— bei Substraten, die in höheren Konzentrationen toxisch sind
— bei Notwendigkeit der exakten Erfassung der jeweiligen Wachstumsphase
— Ermittlung der maximalen Verdünnungsrate (Wachstumsrate)

4.2.1. Kontinuierliche Kultur

Eine chemostatisch geführte Mikroorganismenkultur [118] ist im wesentlichen ein „trial and error"-Experiment. Wird die Substratzufuhr in Abhängigkeit von einem physiologischen Parameter, z. B. dem O_2-Bedarf, durchgeführt, so wird jeweils nur in dem Maße Substrat zugeführt als es auch verarbeitet werden kann, und die Auswaschung der Kultur wird vermieden.
Das prinzipielle Kontrollkonzept ist im folgenden Flußdiagramm dargestellt (Abb. 4).
Die Steuerung des Substratzulaufes erfolgte anfänglich nach einer einfachen Ein-Aus-Regelung, was zu stoßartiger Substratzufuhr führte. In der Folge wurde aus der jeweiligen Förderleistung (DA) durch Mittelwertbildung die „Sollwert-Förderleistung" (DP) errechnet. Falls der O_2-Verbrauch (OA) unter dem Sollwert (OS) liegt, läuft die Pumpe mit folgender Geschwindigkeit:

$$DA = DP + (OS - OA) \cdot K \qquad (K = \text{Konstante})$$

Überschreitet der O_2-Verbrauch der Kultur den Sollwert, so wird die Pumpe gestoppt.

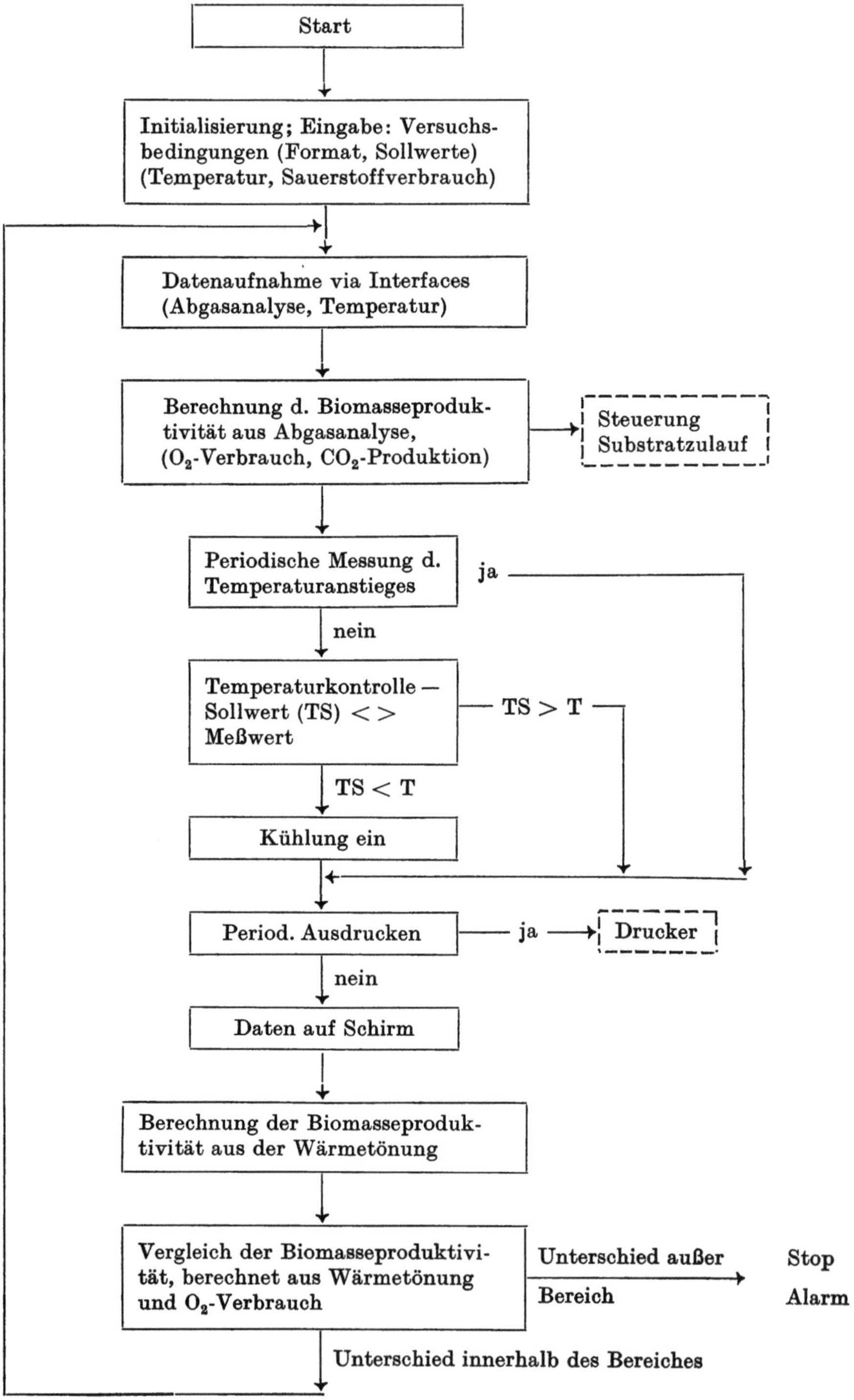

Abb. 4 Flußdiagramm zur Steuerung kontinuierlicher Kulturen von *C. intermedia*

Bei der Ein/Aus-Regelung lagen die Schwankungen bei ± 1,3% absolut; durch obigen Algorithmus konnte eine Glättung dieser Extremwerte erzielt werden und zwar von ± 0,2% absolut.

4.2.2. Steuerung von Zulauf-Fermentationen (fed-batch)

In einer fed-batch- oder Zulauf-Fermentation wird das Substrat stufenweise, entsprechend dem Wachstum nach kinetischen Funktionen vom Rechner gesteuert, zugesetzt. Für den konzeptionellen Aufbau einer Zulaufsteuerung ist die Statuserfassung der Kultur und die Implementierung von kinetischen Funktionen notwendig. Das Konzept der Substratzulaufsteuerung entspricht im wesentlichen einer feed-foreward-Steuerung, wie sie beispielsweise von PERINGER und BLACHERE [15] und auch von WANG et al. [22] für Backhefefermentationen entwickelt wurden.
Zur Demonstration der erarbeiteten Algorithmen wurde eine fed-batch- und eine batch-Kultur von *C. intermedia* betrieben. Die Ausbeute der fed-batch-Kultur lag mit 44,8% deutlich über der der batch-Kultur (39,5%). Die aus dem O_2-Verbrauch errechnete Biomasse der batch-Kultur weicht in der 4. bis 7. Stunde von der experimentell bestimmten geringfügig ab. Die Endwerte stimmen wiederum gut überein. Bessere Übereinstimmung zeigten die Werte der fed-batch-Kultur. Die Volumenentwicklung läuft praktisch parallel mit dem Wachstum, d. h. die Steuerung des Substratzulaufes war adäquat.

4.2.3. Fermentationen mit Rückführung der Biomasse

Fermentationen mit Rückführung der Biomasse, wobei die Biomasse durch Sedimentation konzentriert wird, wurden am Institut für Angewandte Mikrobiologie (Universität für Bodenkultur, Wien) traditionell über lange Zeit betrieben [119, 120].
Derartige Systeme können auch als Modelle für Belebtschlammanlagen betrachtet werden, an denen grundlegende kinetische Untersuchungen durchgeführt werden können. Das System stellt vom Standpunkt der mikrobiologischen Prozeßkontrolle ein wenig eine Herausforderung dar. In letzter Zeit wird es vermehrt in der Praxis eingesetzt, z. B. in der Biospritproduktion [121—123] und in der Verhefung von Sulfitablauge [124].
Einen Schlüsselparameter derartiger Systeme stellt die Sedimentation dar. Ist diese ungenügend, werden in den Fermentor nicht mehr genügend „Reaktionspartner" zurückgeführt. In der Folge schwimmen die Mikroorganismen auf. Aus Vorversuchen war bekannt, daß mit rein physikalischen Meßmethoden (z. B. Messung der optischen Dichte oder Kapazitätsmessung) die Menge an sedimentierter Biomasse nicht ausreichend charakterisiert werden kann. Interessanter wäre neben der rein mengenmäßigen Messung der sedimentierten Biomasse die Aktivität derselben.
Die Flockulation scheint gemäß Literaturbefunden [125—127] durch von den Mikroorganismen ausgeschiedene Polymere, die an den aktiven Zentren der Zelloberfläche absorbiert werden, bewirkt zu werden. Durch diese Polymere kommt es zu Brückenbildungen zwischen den einzelnen Zellen. Dies bestätigen auch Arbeiten von SMIDT [128]. Durch Zusatz bestimmter Proteinfraktionen aus Bierwürze konnte die Flockulation induziert werden. Darüberhinaus stellte STEWART [129] fest, daß die Fähigkeit zur Flokkulation genetisch determiniert ist.
Im folgenden Kontrollkonzept [130, 131] wurde die Steuerung der Biomasserückfuhr nach Substratpulsen nach zwei Arbeitsweisen vorgenommen:

- Steuerung der Biomasse-Rückfuhr proportional zur Zulaufrate (1/3—1/4)
- Steuerung der Biomasse-Rückfuhr proportional-differential zur Zulaufrate (PA = 120/E, PA = Pumpenleistung, E = Sollwertabweichung).

Zusätzlich enthält dieser Algorithmus Funktionen, um Störungen des Gleichgewichtes, die von außen an das System herangetragen werden, erkennen zu können. Massive

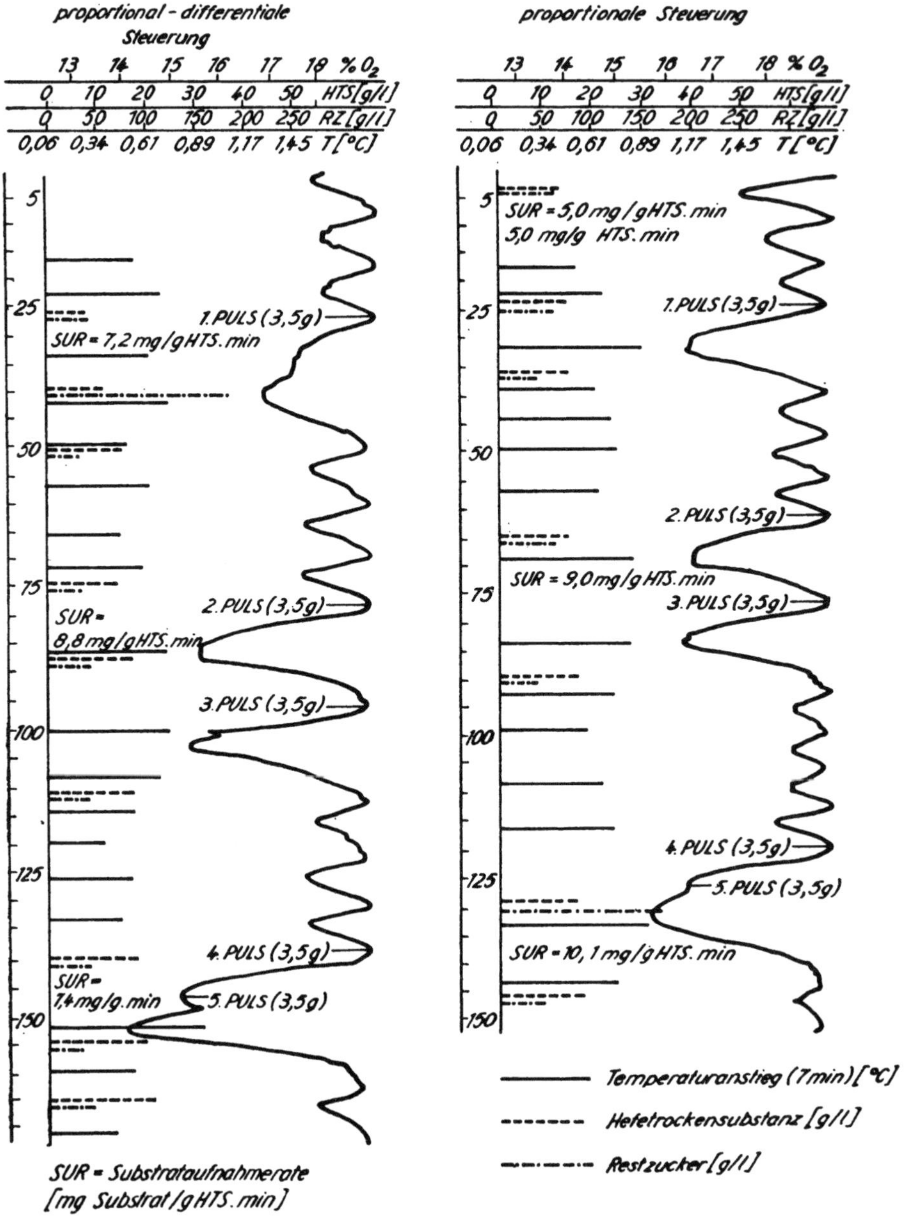

Abb. 5. Kontinuierliche Kultur von *C. intermedia* mit Rückführung der Biomasse.

Störungen, die die Aktivierung des differentialen Anteils der Rückführsteuerung notwendig machen, werden durch signifikante Abweichungen des O_2-Gehaltes vom Sollwert und durch ständige Zunahme des O_2-Verbrauches erkannt.

Bei konstantem Substratzulauf kann mit den für die Steuerung einer kontinuierlichen Kultur erarbeiteten Algorithmen und der proportionalen Rückfuhrregelung der sedimentierten Biomasse der Gleichgewichtszustand aufrechterhalten werden. Die Bio-

masseproduktivität nimmt gleichmäßig zu. Im gleichen Maße steigt die HTS im Fermentor (gemäß der proportionalen Rückführung). Es ist keine aufwendige Regelung notwendig. Um die Dynamik eines derartigen Systems sichtbar zu machen, bedient man sich wiederum der Puls-Technik.
Mittels der proportional differentialen Regelung soll die Störung erkannt werden und die Biomasserückfuhr gesteuert werden. Wie aus Abb. 5 ersichtlich ist, ist die proportional differentiale Regelung in der Lage, nach dem 1. Puls die Störung zu erkennen und vermehrt Biomasse rückzuführen. Der Restzuckergehalt, enzymatisch bestimmt (Boehringer Test, Kombination Saccharose/Glucose), liegt nach allen folgenden Pulsen im Bereich von 50—70 mg Saccharose/l. Anders liegen die Verhältnisse bei der proportionalen Regelung: der Rest-Saccharosegehalt steigt ständig an, die Hefe zeigt zwar die maximale Substrataufnahmerate, aber die Rückführung ist ungenügend, um die Saccharose vollständig abzubauen. Diese hohen spezifischen Substrataufnahmeraten und auch Umsatzraten äußern sich auch in den kalorimetrischen Werten. Die proportional differentiale Steuerung zeigt in den Perioden nach den Pulsen ein deutliches Absinken der Wärmetönung im Gegensatz zur rein proportionalen.
Mittels des Sensorsystems Atmung können Störungen rasch erkannt und geeignete Strategien zu deren Behebung aktiviert werden. Mit einer ähnlichen Problematik beschäftigten sich Gold et al. [124]. Sie führten Computer-kontrollierte Fermentationen mit Rückführung der Biomasse durch. Die Problemstellung war die Verhefung von Sulfitablauge (*C. utilis*). Die Sedimentation wurde mit Bentonit stabilisiert. Die Biomasse im Fermentor wurde aus der Trübung bestimmt, die übrigen Parameter mittels Komponentenbilanzierung. Die Rückführung der Biomasse erfolgte nur proportional zum Zulauf (0,55 h^{-1}). Durch die Rückführung konnte die Biomasseproduktivität um 150% gesteigert werden.

Die experimentellen Arbeiten wurden in dankenswerter Weise vom Fonds zur Förderung der wissenschaftlichen Forschung (Projekt 3895) finanziell unterstützt.

Eingegangen: 19. 5. 1983

Literatur

[1] Blanch, H. W., Dunn, I. J.: Adv. Biochem. Eng. **3** (1974), 127. Hrsg. Ghose, T. K., Fiechter, A. und Blakebrough, N., Heidelberg, New York: Springer-Verlag.
[2] Roels, J. A., Kossen, N. W. F. In: Progress in Industrial Microbiology. Ed. Bull, M. J., Amsterdam, Oxford, New York: Elsevier Scientific Publ. Comp. 1978, S. 95.
[3] Howell, J. A., Jones, M. G.: Symposium: Computer Application in Fermentation Technology, Manchester, GB.
[4] Chase, L. M.: Biotechnol. Bioeng. **19** (1977), 1431.
[5] Holmberg, A., Ranta, J.: Automatica **18** (1982) 2, 181.
[6] Therien, N., Harrington, P.: Proc. 11th Canadian Symp. Water Pollution Research, Canada, 1976.
[7] Monod, J.: Recherches sur la croissance des cultures bacteriennes. Herman et Cie, Paris, 1942.
[8] Teissier, G.: Ann. Physiol. Physicochim. Biol. **12** (1936).
[9] Moser, H.: Carnegie Institute Washington, 1958. Pub. Nr. 614.
[10] Mikesell, R. D.: J. Sanit. Eng. Div. Proc. Am. Soc. Civil Engrs. 1971.
[11] Hollo, J., Donath-Jobbagy, A.: Biotechnol. Lett. **2** (1980) 2, 55.
[12] Clifft, R. C., Andrews, J. F.: J. Wat. Poll. Cont. Fed. **53** (1981) 7, 1219.
[13] Contois, D. E.: J. Gen. Microbiol. **21** (1959), 40.
[14] Chen, Y. R., Hashimoto, A. G.: Biotechnol. Bioeng. **22** (1980), 2081.
[15] Peringer, P., Blachere, H. T.: Biotechnol. Bioeng. Symp. **9** (1979), 205.
[16] Bajpai, R. K., Reuss, M.: Biotechnol. Bioeng. **23** (1981), 717.

[17] CONSTANTINIDES, A., SPENCER, J. L., GADEN jr., E. L.: Biotechnol. Bioeng. **12** (1970), 803.
[18] BLANCH, H. W., ROGERS, P. L.: Biotechnol. Bioeng. **13** (1971), 843.
[19] CHING, D. L., MAIER, J. W.: 10th Ann. Pittsburgh Conf. on Modelling and Simulation, Pittsburgh, PA. 1979, April, S. 25–27.
[20] MEISTER, G.: Diplomarbeit, Universität für Bodenkultur, Wien, 1981.
[21] BAYER, K., MEISTER, G.: Eur. J. Appl. Microbiol. Biotechnol. **15** (1982), 36.
[22] WANG, H. Y., COONEY, C. L., WANG, D. I. C.: Biotechnol. Bioeng. **21** (1979), 975.
[23] SPRUYTENBURG, R., DUNN, I. J., BOURNE, J. R.: Biotechnol. Bioeng. Symp. **9** (1979), 359.
[24] KNORRE, W. A., GUTHKE, R. In: Continuous Cultivation of Microorganisms. Proceedings of the 7th Symposium in Prague. Eds. Sikyta, B., Fencl, Z. und Polacek, V., Institute of Microbiology, Czechoslovak Academy of Sciences, Prague. 1980, S. 199.
[25] MUZYCHENKO, L. A., MASCHEVA, L. A., YAKLOVELVA, G. V.: Biotechnol. Bioeng. Symp. **4** (1974), 629.
[26] JERUSALEMSKY, N. D. In: Microbial Physiology and Continuous Culture. Eds. Powell, E. O., Evans, C. G. T., Strange, R. E., Tempest, D. W., Her Majesty's Stationary Office (HMSO). 1967, S. 23.
[27] PONTRYAGIN, S. L., BOLTYANSKII, L. S., GAMKRELIDZE, R. V., MISCHENKO, E. F.: The Mathematical Theory of Optimal Process (Engl. Übers. von K. N. Trirogoff), New York: Wiley Interscience. 1962.
[28] FAN, L. T.: The Continuous Maximum Principle. New York: Wiley Interscience, 1966.
[29] UNDEN, G. A., HEDEN, C. G.: The First European Conference on Computer Process Control in Fermentation. Dijon (France), 1973.
[30] CORRIEU, G., LALANDE, M., PERINGER, P., JARROT, C., ALMANZA, R.: Rev. Ferment. Ind. Aliment., Bruxelles, **30** (1976) n 5, 125.
[31] NYESTE, L., VERES, A., SZIGETI, L., KURUCZ, I., PUNGOR, E., HOLLO, J.: Abh. Akad. Wissensch. DDR, Abt. Math. Naturwissensch. Techn. 3N (1979) 347.
[32] NYESTE, L., SZIGETI, L., VERES, A., PUNGOR, E., KURUCZ, I., HOLLO, J.: Biotechnol. Bioeng. **23** (1981), 405.
[33] VERES, A., NYESTE, I., KURUCZ, L., KIRCHKNOPF, L., SZIGETI, L., HOLLO, J.: Biotechnol. Bioeng. **23** (1981), 391.
[34] NELDER, J. A., MEAD, R.: Computer J. **7** (1965), 308.
[35] SWANN, W. H.: FEBS-Letters **2**, Supplement (1969) March, 39.
[36] POWELL, M. J. D.: Computer J. **7** (1964), 155.
[37] HOOKE, R., JEEVES, T. A.: Journal of the Association for Computing Machinery 8 (1961), 212.
[38] BOX, G. E. P., WILSON, K. B.: J. Roy. Stat. Soc. **13** (1951), 1.
[39] SCHRÖDER, K. D., WEIDE, H.: Biotechnol. Bioeng. Symp. **4** (1974), 713.
[40] REMMERS, E. G.: Diss. MIT, Cambridge, Mass.
[41] CONSTANTINIDES, A., SPENCER, J. L., GADEN jr., E. L.: Biotechnol. Bioeng. 12 (1970), 1081.
[42] MARTEN, J. In: Methods in Microbiology. Eds. Norris, J. R. and Ribbons, D. W., London, New York: Academic Press, 1972.
[43] NESTAAS, E., WANG, D. I. C.: Adv. Biotechnol. **1** (1980), 433. Eds. Mou-Young, M., Robinson, C. W., Vezina, C., Toronto, Oxford, New York: Pergamon Press.
[44] YANO, T., KOBAYASHI, T., SHIMIZU, S.: J. Ferment. Technol. **56** (1978) 4, 421.
[45] TANNEN, L. P., NYIRI, L. K. In: Microbial. Technology — Fermentation Technology, Vol. II. Eds. Peppler, H. J. and Perlmann, D., New York: Academic Press, 1979.
[46] RIBITSCH, V.: Persönliche Mitteilung. Wissenschaftsmesse Wien, 1981.
[47] KRATKY, O., LEOPOLD, H., STABINGER, H.: Z. Angew. Phys. **4** (1969), 273.
[48] CAMMAN, K.: Berlin, Heidelberg: Springer-Verlag, 1979.
[49] WIMPENNY, J. W. T.: J. Appl. Chem. Biotechnol. **26** (1969), 48.
[50] BUCKLAND, R. C., FASTERT, H.: Symposium: Computer Application in Fermentation Technology, Manchester, GB. 1981.
[51] FIECHTER, A., MEISNERS, M., SUKATSCH, D. A.: Biologische Regulation und Prozeßführung. In: Handbuch der Biotechnologie. Hrsg. Präve, Faust, Sittig und Sukatsch. Wiesbaden: Akademische Verlagsgesellschaft, 1982, S. 173–215.
[52] BACH, H. P., WÖHRER, W., RÖHR, M.: Biotechnol. Bioeng. **20** (1978), 799.
[53] BEYELER, W., EINSELE, A., FIECHTER, A.: Eur. J. Appl. Microbiol. Biotechnol. **13** (1981), 10.

[54] McElroy, W. D.: Proc. Natl. Acad. Sci. USA **33** (1947), 432.
[55] Levin, G. V., Clendenning, J. R., Chapelle, E. W., Heim, A. H., Rocek, E.: Bioscience **14** (1964), 37.
[56] Thomanetz, E., Sperandio, A., Bardtke, D.: GWF-Wasser/Abwasser **123** (1982), 96.
[57] Thomanetz, E., Sperandio, E., Bardtke, D.: GWF-Wasser/Abwasser **123** (1982) 2, 147.
[58] Bowers, L. D., Carr, P. W.: Biochemical Engineering **15** (1980), 89. Hrsg. Ghose, T. K., Fiechter, A. und Blakebrough, N., Berlin, Heidelberg, New York: Springer-Verlag.
[59] Mor, J. R.: Symposium: Computer Application in Fermentation Technology, Manchester, GB. 1981.
[60] Suzuki, S., Karube, I.: Appl. Biochem. Bioeng. **3** (1981), 145.
[61] Suzuki, S., Sato, I., Karube, I.: Appl. Biochem. Biotechnol. **7** (1982), 147.
[62] Danielsson, B., Mosbach, K.: Symposium: Computer Application in Fermentation Technology, Manchester, GB. 1981.
[63] Schultz, J. S., Sims, G.: Biotechnol. Bioeng. Symp. **9** (1979), 65.
[64] Zabriskie, D. W.: Annals of New York Academy of Science, 1979. S. 223.
[65] Karl, D. M.: Microbial Revies **44** (1980) 4, 739.
[66] Karube, I., Matsunaga, T., Mitsuda, S., Suzuki, S.: Biotechnol. Bioeng. **19** (1977), 1535.
[67] Hikuma, M., Suzuki, H., Yasuda, T., Karube, I., Suzuki, S.: Eur. J. Appl. Microbiol. Biotechnol. **8** (1979), 289.
[68] Kobos, R. K., Pyon, H. Y.: Biotechnol. Bioeng. **23** (1981), 627.
[69] Danielsson, B., Mattiasson, B., Mosbach, K.: Appl. Biochem. Bioeng. **3** (1981), 97. London, New York: Academic Press.
[70] Cooney, C. L., Wang, H. Y., Wang, D. I. C.: Biotechnol. Bioeng. **19** (1977), 55.
[71] Zabriskie, D. W., Armiger, W. B., Humphrey, A. E.: Workshop Computer Applied in Fermentation Technology GBF. Hrsg. Jeffries III R. P., Weinheim: Verlag Chemie. 1976, S. 59.
[72] Harrison, D. E. F.: J. Appl. Chem. Biotechnol. **22** (1972), 417.
[73] Nyiri, K. L., Jeffries, R. P., Humphrey, A. E.: Biotechnol. Bioeng. Symp. **4** (1974), 613.
[74] Humphrey, A. E.: The First European Conference on Computer Process Control in Fermentation, Dijon (France), 1973.
[75] Wang, H. Y., Cooney, S. L., Wang, D. I. C.: Biotechnol. Bioeng. **19** (1977), 69.
[76] Swartz, J. R., Cooney, C. L.: Biotechnol. Bioeng. Symp. **9** (1979), 95.
[77] Wang, H. Y., Cooney, C. L., Wang, D. I. C.: Biotechnol. Bioeng. Symp. **9** (1979), 13.
[78] Nanba, A., Hirota, F., Nagai, S.: J. Ferment. Technol. **59** (1981) 5, 383.
[79] Hampel, W., Wöhrer, W., Bach, H. P., Röhr, M.: Fifth International Fermentation Symposium, Berlin, 1976.
[80] Hampel, W., Bach, H. P., Röhr, M.: Workshop Computer Applied in Fermentation Technology GBF. Hrsg. Jeffries, III, R. P., Weinheim, New York: Verlag Chemie, 1977.
[81] Humphrey, A. E.: Proceedings of Labex Symposium, Earls Court, London, 1971, April.
[82] Weilbuchner, J.: Diplomarbeit, Universität für Bodenkultur, Wien, 1977.
[83] Führer, F.: Diplomarbeit, Universität für Bodenkultur, Wien. 1982.
[84] Bravard, J. P., Cordonnier, M., Kernevez, J. P., Lebault, J. M.: Biotechnol. Bioeng. **21** (1979), 1239.
[85] Bayer, K.: Diss., Universität für Bodenkultur, Wien. 1977.
[86] Fiechter, A., v. Meyenburg, K.: Biotechnol. Bioeng. **10** (1968), 535.
[87] Sols, A., de la Fruente, G.: Proc. Membrane Transport and Metabolism, Prague. 1960, S. 361–377.
[88] Leegwater, M. P. M., Neijssel, O. M., Tempest, D. W.: J. Chem. Technol. Biotechnol. **32** (1982), 92.
[89] Tempest, H.: J. Gen. Microbiol. **41** (1965), 143.
[90] Forrest, W. W., Walkner, D. J. In: Advances in Microbial Physiology. Eds. Rose, Wilkinson, London, New York: Acad. Press. 1971.
[91] Cooney, S. L., Wang, D. I. C., Mateles, R. I.: Biotechnol. Bioeng. **11** (1968), 269.
[92] Mou, D. G., Cooney, C. L.: Biotechnol. Bioeng. **18** (1976), 1371.
[93] Wang, H., Wang, D. I. C., Cooney, C. L.: Eur. J. Appl. Microbiol. Biotechnol. **5** (1978), 207.
[94] Fortier, J. L., Reboul, B., Philip, P., Simard, M. A., Picker, P., Jolicoeur, C.: J. Water Pollution Control Federation **52** (1980) 1, 89.
[95] Plasse, F., Fardeau, M. L., Belaich, J. P.: Biotechnol. Lett. **2** (1980) 1, 11.

[96] SMITH, R. J.: Symposium: Computer Application in Fermentation Technology, Manchester, GB. 1981.
[97] SWARTZ, J., WANG, H., COONEY, C. L., WANG, D. I. C.: Fifth International Fermentation Symposium, Berlin, 1976.
[98] BLACHERE, H. T.: The 1th Europ. Conference on Computer Process Control in Fermentation, Dijon, (France), 1973.
[99] COONEY, C. L.: Biotechnol. Bioeng. Symp. **9** (1979), 1.
[100] FALCH, E. A., HJORTKJAER, P., PEDERSEN, P.: Symposium: Computer Application in Fermentation Technology, Manchester, GB. 1981.
[101] LUNDELL, R.: Symposium: Computer Application in Fermentation Technology, Manchester, GB. 1981.
[102] ALFORD, J. S.: Symposium: Computer Application in Fermentation Technology, Manchester, GB. 1981.
[103] AIBA, S.: Biotechnol. Bioeng. Symp. **9** (1979), 269.
[104] DAIRAKU, K., YAMASAKI, Y., KUKI, K.: Biotechnol. Bioeng. **23** (1981), 2069.
[105] KALOGERAKIS, N., BOYLE, T. J.: Biotechnol. Bioeng. **23** (1981), 921.
[106] SPRUYTENBURG, R., DUNN, I. J., BOURNE, R. J.: 1th Europ. Congr. on Biotechnol., Interlaken. 1978.
[107] WÖHRER, W., HAMPEL, W., RÖHR, M.: Adv. Biotechnol. **1** (1980), 419. Eds. Moo-Young, M., Robinson, C. W., Vezina, C., Toronto, Oxford, New York: Pergamon Press, 1980.
[108] HALME, A.: Biotechnol. Bioeng. Symp. **9** (1979), 369.
[109] MOILANEN, U. R.: Symposium: Control Science and Technology for the Progress of Society. International Federation of Automatic Control, 8th Triennal World Congress. Kyoto (Japan). August, 1981.
[110] TONGE, G. M.: Symposium: Computer Application in Fermentation Technology, GB. 1981.
[111] KUBOTA, H., FUJIE, K., KASAKURA, T.: Water Sci. Technol. **13** (1981), 159.
[112] BERNARD, A., CORDONNIER, M., LEBEAULT, J. M.: Adv. Biotechnol. **1** (1980), 407. Eds. Mou-Young, M., Robinson, C. W., Vezina, C., Toronto, Oxford, New York: Pergamon Press.
[113] SPÄTH, H.: Spline-Algorithmen zur Konstruktion glatter Kurven und Flächen. München, Wien: R. Oldenburg Verlag. 1978.
[114] JEFFRIS, R. P., WINTER, H., VOGELMAN, H.: Workshop Computer Application in Fermentation Technology. Hrsg. Jeffris III, R. P., Weinheim, New York: Verlag Chemie, 1976.
[115] SVERCEK, W. Y., ELLIOT, R. F., ZAJIC, J. E.: Biotechnol. Bioeng. **16** (1974), 827.
[116] JEFFRIS, R. P.: Computer Application in Fermentation Technology, Manchester, GB. 1981.
[117] SELVA, G., ZERBETTO, P.: Workshop Computer Application in Fermentation Technology GBF. Hrsg. Jeffris III, R. P., Weinheim, New York: Verlag Chemie, 1977.
[118] NOVICK, A., SZILARD, L.: Science **112** (1950), 715. NY.
[119] MEYRATH, J.: 4th Int. Symp. on Yeasts, Vienna, 1974.
[120] BRAUN, R., MEYRATH, J.: Congres International Microbiologie et Industrie Alimentaire, APRIA, Paris, 1979, Oktober.
[121] MEYRATH, J.: Congres International Microbiologie et Industrie Alimentaire, APRIA, Paris. 1979, Oktober.
[122] BRÜSCHKE, H., TUSEL, G. F., BALLWEG, A. H.: Firmendokumentation, INCON-Anlagentechnik, D-6650 Hamburg. 1979.
[123] WEEKS, M. G., MUNRO, P. A., SPEDDING, P. L.: Biotechnol. Lett. **4** (1982) 2, 85.
[124] GOLD, D., MOHAGHEGHI, A., COONEY, C. L., WANG, D. I. C.: Biotechnol. Bioeng. **23** (1981), 2105.
[125] FRIEDMANN, B. A., DUGAN, P. R., PFISTER, R. M., BENSON, C. R.: J. Bacteriol. **98** (1969), 1328.
[126] PAVONI, J. L., TENNEY, M. W., ECHELBERGER jr., W. F.: J. Water Pollution Control Federation **4** (1979) 51, 798.
[127] GULAS, V., BOND, M., BENEFIELD, L.: J. Water Pollution Control Federation **4** (1979) 51, 798.
[128] SMITH, E.: Diplomarbeit, Universität für Bodenkultur, Wien. 1976.
[129] STEWART, G. G., RUSSEL, F.: Canad. J. Microbiol. **23** (1977) 4, 441.
[130] MACHEINER, A.: Diplomarbeit, Universität für Bodenkultur, Wien, 1982, in press.
[131] BAYER, K., MACHEINER, A.: Jahrestagung der ÖGHMP, Millstatt, 1982.

Acta Biotechnol. 4 (1984) 3, 235—245

Vergleich von Messungen des $k_L a$-Wertes nach der stationären und der dynamischen Methode bei der Fermentation von L-Lysin und Turimycin

H.-O. MÖCKEL

VEB Komplette Chemieanlagen Dresden im
VEB Chemieanlagenbaukombinat Leipzig/Grimma
8012 Dresden, PSF 184

Summary

The calculation and scale-up of fermentation processes need $k_L a$ as one of the most important engineering data. There are two methods to determine $k_L a$ depending on power input, aeration rate and the properties of the fermentation broth: static with a balance between air supply and exit, dynamic gassing out with following the changes of dissolved oxygen concentration during periods of air off and a following air on.
Within early intervals of fermentation time the data from both methods agree well, while for later time intervals the dynamic method always gives much lower values for $k_L a$ than static. The only explanations for this phenomenon are quick changes in the oxygen metabolism or an enzymatic storage of oxygen. For both gassing out and saturation period it is possible to calculate the same absolute amounts of this additional oxygen.

Für die Modellierung von aeroben Fermentationsprozessen stellt der $k_L a$-Wert die wichtigste Kenngröße dar, die von geometrischen und verfahrenstechnischen Parametern sowie von den durch das biologische System bestimmten Stoffeigenschaften abhängig ist. Grundsätzlich sind zwei Wege zur Messung des $k_L a$-Wertes bekannt:

1. Stationäre Methode

Aufstellen der Sauerstoffbilanz über den Fermentor durch Analyse von Zu- und Abluft und Messung der Gelöstsauerstoffkonzentration, Berechnung der Sauerstoffübergangsrate und des $k_L a$-Wertes.

2. Dynamische Methode

Zeitweilige Unterbrechung der Belüftung. Aus der Auszehrungsgeschwindigkeit des gelösten Sauerstoffes kann die Sauerstoffübergangsrate, aus der Aufsättigung nach Wiedereinschalten der Begasung der $k_L a$-Wert berechnet werden.
Der Gelöstsauerstoffspiegel wird dazu mit einer schnellansprechenden Sauerstoffelektrode verfolgt.
Es war zu untersuchen, inwieweit beide Methoden übereinstimmende Ergebnisse liefern.

Bisherige Arbeiten

Die genannten Meßmethoden sind in der Literatur seit langem bekannt, wie eine ausführliche Übersicht [1] verdeutlicht. Da die stationäre Methode bei größerem apparativen Aufwand meßtechnisch geringere Schwierigkeiten bereitet, sind ihr nur wenige

Arbeiten gewidmet [2]. Die dynamische Methode ist demgegenüber mit einer Reihe meßtechnischer Probleme behaftet, die durch die Diffusionswiderstände von Membran und Grenzschicht [3—8], durch das an der Elektrode herrschende örtliche Gas-hold-up [7—10], durch die dynamischen Effekte beim hold-up-Aufbau und dem sich ändernden Sauerstoffpartialdruck der Gasphase während der Aufsättigung [11, 12] entstehen.
Diese theoretischen Arbeiten, die meist mit einer Überprüfung im Modellmedium verbunden sind, müssen durch Untersuchungen mit originalen Fermentationssystemen ergänzt werden.
In [13] wird über erste Erfahrungen berichtet, wobei als Fehlerquellen der Lufteintrag durch die bewegte Oberfläche und die bei länger ausgesetzter Belüftung sich verändernde Atmungsrate genannt werden.
Die Auswertung der Verläufe von Auszehrung und Aufsättigung in [14] wird durch Auswahl charakteristischer Punkte ausgeführt. Die so bestimmten k_La-Werte stimmten gut mit den nach [13] ermittelten überein. Der Einsatz der dynamischen Methode in einer Kolonne bei Verwendung eines Modellabwassers wird in [15] dargestellt. Die in [13] beschriebene Auswertemethode wird auch in [16] verwendet.
Es fällt auf, daß in keiner der zuletzt genannten Arbeiten der k_La-Wert korrigiert wird, um die durch Membranwiderstände und dynamische Effekte entstehenden Fehler auszugleichen. Ein Weg, die Anwendbarkeit der dynamischen Methode in Fermentationsmedien zu überprüfen, besteht darin, gleichzeitig stationär und dynamisch zu messen und damit Sauerstoffübergangsrate r_{O_2} und k_La-Wert unabhängig voneinander zu bestimmen.

Experimentelles

Die Untersuchungen wurden an zwei verschiedenen Kulturen vorgenommen:

1. Turimycinbildner
Streptomyces hygroscopicus
Nährboden und Fermentationsablauf sind in [17] beschrieben.

2. Lysinbildner
Corynebacterium glutamicum
Nährboden und Fermentationsablauf sind in [18] beschrieben.

Es stand ein Fermentor mit einem Füllvolumen von $V_L = 0{,}058\ m^3$ zur Verfügung, Behälterdurchmesser 0,4 m, Füllhöhe 0,5 m, 6-Blatt-Schaufelrührer mit 0,16 m Durchmesser. Die Rührerdrehzahl war stufenlos regelbar. Durch Drehmomentmessung konnte die Rührernettoleistung P_R bestimmt werden. Die zugeführte Luftmenge wurde mittels Rotameter gemessen.
Die Sauerstoffelektrode war so angebracht, daß sie vom Förderstrom des Rührers direkt angeströmt wurde, um das Anhaften von Gasblasen und Grenzschichteffekte zu minimieren. Die Sondenzeitkonstante betrug $\tau_S = 4$ s. Über einen Meßverstärker wurde das Sondensignal auf einem Schreiber mit variabler Vorschubgeschwindigkeit registriert.
Für die Abluftanalyse wurden für den Sauerstoffpartialdruck $\psi_{O_{2A}}$ ein Permolyt und den CO_2-Partialdruck $\psi_{CO_{2A}}$ ein Infralyt in Verbindung mit Schreibern eingesetzt.

Versuchsauswertung

Es werden aus den Meßwerten berechnet:

1. Stationäre Methode

$$r_{O_2 \text{stat}} = \varrho_{O_{2N}} \frac{\dot{V}_{GN}}{V_L} \frac{\psi_{O_{2E}} - \psi_{O_{2A}} - \psi_{O_{2E}} \cdot \psi_{CO_{2A}}}{1 - \psi_{O_{2A}} - \psi_{CO_{2A}}}$$

$$k_L a_{\text{stat}} = \frac{r_{O_2 \text{stat}}}{c^*_{O_{2A}} - c_{L\infty}} \quad \text{mit} \quad c^*_{O_{2A}} = H \cdot \overline{p} \cdot \psi_{O_{2A}}$$

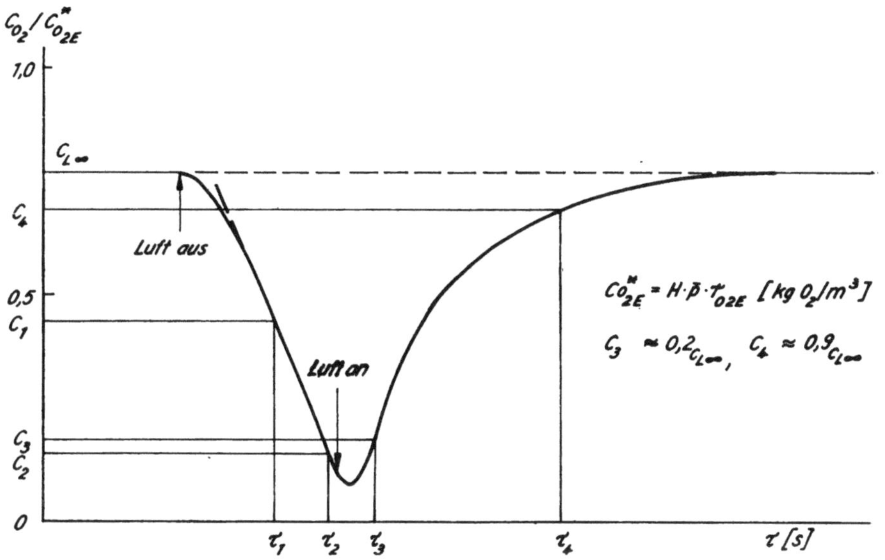

Abb. 1. Schematischer Verlauf von Auszehrung und Aufsättigung

2. Dynamische Methode

Aus dem Verlauf des relativen Gelöst-O_2-Spiegels entsprechend Abb. 1 werden mit den angegebenen charakteristischen Kurvenpunkten berechnet:

$$r_{O_2 \text{dyn}} = \frac{c^*_{O_{2E}}(c_1 - c_2)}{\tau_2 - \tau_1} = \frac{dc_L}{d\tau} \quad \text{mit} \quad c^*_{O_{2E}} = H \cdot \overline{p} \cdot \psi_{O_{2E}}$$

Für den Fall, daß der lineare Bereich der Auszehrungskurve sehr kurz ist (schäumende, feinblasige Dispersion), kann nur der Kurvenabschnitt unmittelbar vor Wiederbeginn der Luftzufuhr ausgewertet werden.

$$k_L a_0 = \frac{\ln \dfrac{1 - c_3}{1 - c_4}}{\tau_4 - \tau_3}$$

Der Wert k_La_O ist wegen der eingangs geschilderten verfahrensbedingten Meßfehler zu korrigieren, wobei die in [19] abgeleiteten Korrekturgrößen verwendet werden:

$$K_G = f\left(k_La_O \cdot \frac{V_G}{\dot{V}_G} \cdot \frac{1-\varphi_G}{\varphi_G} \frac{\bar{p} \cdot H}{\varrho_{O_2}}\right) \quad \text{Korrekturfaktor Gasverweilzeit}$$

$$K_S = f\left(\frac{\tau_S}{\tau_4 - \tau_3}\right) \quad \text{Korrekturfaktor Sondenträgheit}$$

$$k_La_{\text{dyn}} = K_G \cdot K_S \cdot k_La_O$$

Die unabhängig voneinander bestimmten r_{O_2}- und k_La-Werte sind für die nach Fermentationszeit, Luftmenge und Leistungseintrag charakterisierten Meßpunkte miteinander zu vergleichen und zu bewerten.

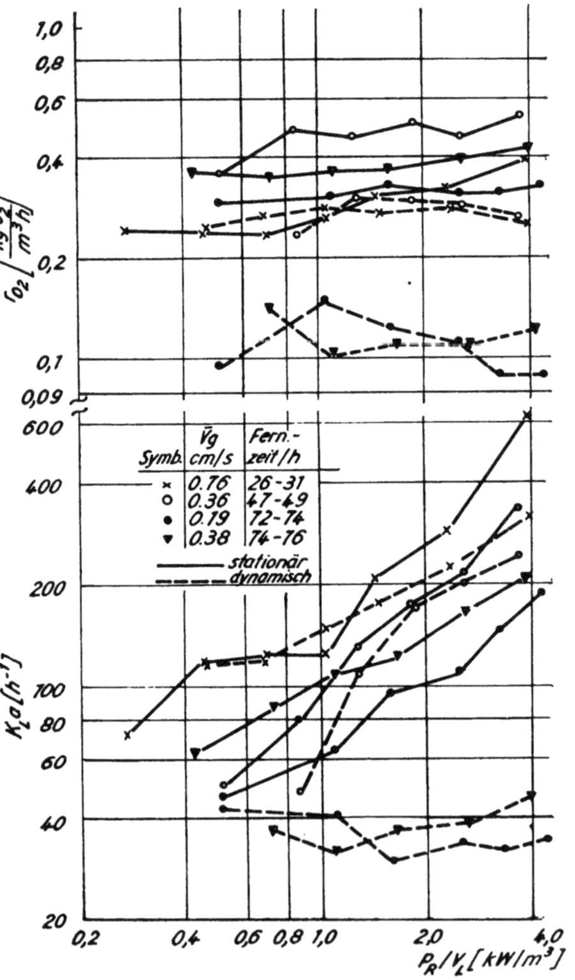

Abb. 2. Vergleich von nach der stationären bzw. dynamischen Methode gemessenen Werten während der Turimycinfermentation I

Ergebnisse

In Abb. 2 sind die auf zwei Wegen ermittelten Kenngrößen r_{O_2} und k_La für die Turimycinfermentation I zu verschiedenen Fermentationszeiten und für variierte verfahrenstechnische Bedingungen miteinander verglichen. Bis zu einer Fermentationszeit von ca. 50 h stimmen die Werte nach beiden Methoden im Rahmen der zu erwartenden Streubreite überein. Bei den Vergleichsmessungen zur 70.—75. Stunde ergeben sich dagegen große systematische Abweichungen dahingehend, daß nach der dynamischen Methode wesentlich kleinere Werte für r_{O_2} und k_La errechnet werden. Die gleiche Aussage läßt sich anhand der Ergebnisse für die Turimycinfermentation II (Abb. 3) treffen. Da alle Messungen mit den gleichen Apparaturen ausgeführt und der Auswerteweg beibehalten wurde, sich weder die Stoffeigenschaften, der Biomasseanteil noch der pH-Wert bedeutend veränderten, kann hier keine Erklärung für die Differenzen gefunden werden. Die Frage, ob der beobachtete Effekt auch bei einer Fermentation anderen Charakters auftritt,

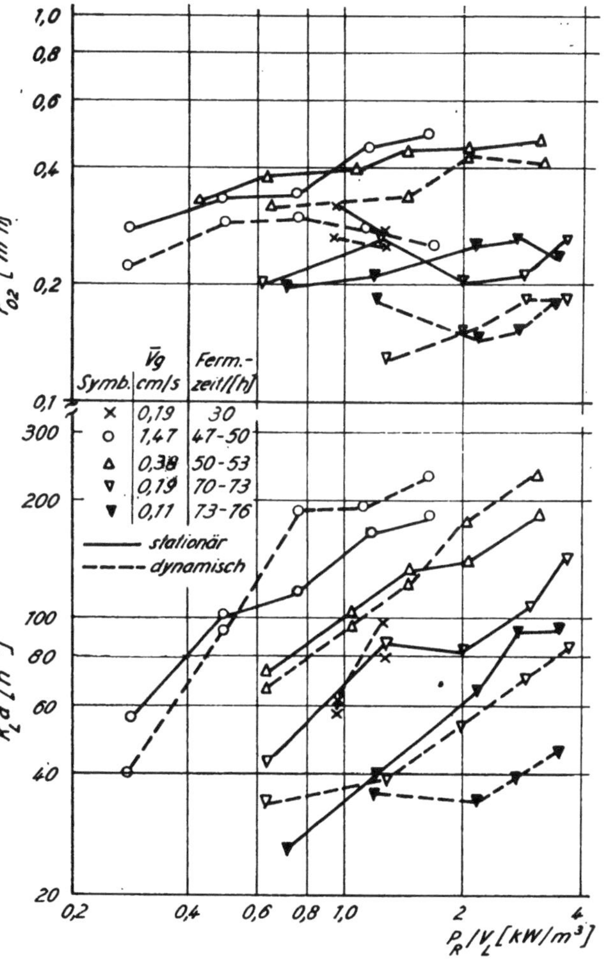

Abb. 3. Vergleich von nach der stationären bzw. dynamischen Methode gemessenen Werten während der Turimycinfermentation II

wurde durch Vergleichsmessungen in einem Lysinfermentationsmedium untersucht. Während der 50.—55. Stunde ausgeführte Messungen (Abb. 4) zeigen ähnliche Unterschiede in den Ergebnissen. Ergänzend ist festzustellen, daß bei Übereinstimmung von stationärer und dynamischer Messung die Auszehrungsgeschwindigkeit $dc_L/d\tau = r_{O_2}$ für $c_L \leqq 0{,}07c^*_{O_2E}$ geringer wird, d. h. $dc_L/d\tau = r_{O_2} = f(c_L)$. Im anderen Fall bleibt $dc_L/d\tau$ bis $c_L = 0$ unverändert. Da sowohl während der Auszehrungs- als auch der Aufsättigungsphase zu niedrige Werte gemessen werden, kann das gefundene Ergebnis nur

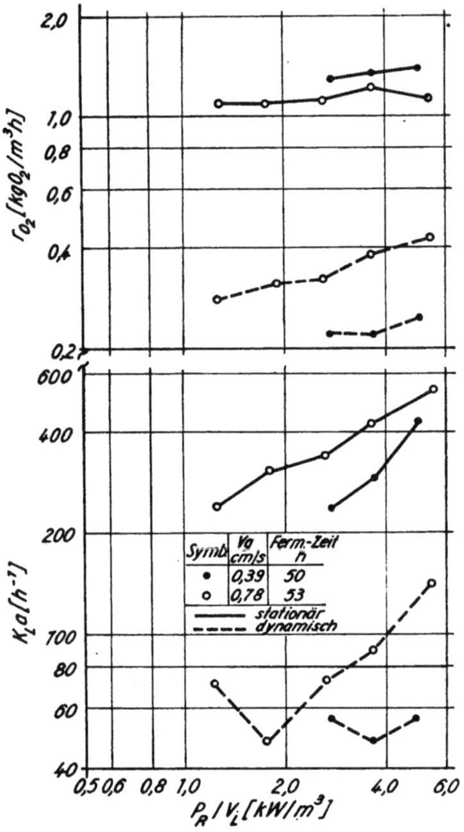

Abb. 4. Vergleich von Messungen nach der dynamischen und nach der stationären Methode für die Lysinfermentation

so erklärt werden, daß durch eine sehr rasche Stoffwechselumschaltung während der Auszehrung zusätzlich zum gelösten Sauerstoff Sauerstoff freigesetzt wird, der während der Aufsättigung zusätzlich wieder angelagert werden muß, da das Ergebnis reproduzierbar ist. Möglich erscheint auch eine enzymatische O_2-Freisetzung bzw. Anlagerung. In [20] wird auf der Basis von kinetischen Versuchen die These aufgestellt, daß in den Mikroorganismen ein labiles Oxidationszwischenprodukt existiert, das Sauerstoff speichern kann. Dabei läuft die Hinreaktion (Oxidation) sehr schnell ab, während die Rückreaktion (O_2-Freisetzung) sehr langsam abläuft, wodurch die Langzeitspeicherwirkung erklärbar ist. Ausgehend von den Angaben in [20] wird in [21] der Verlauf von Auszehrungs- und Aufsättigungskurve nachgerechnet. Eine rein physikalische Speicherung innerhalb der Biomasse ist nicht möglich, da höhere Konzentrationen als in der umgebenden Flüssigkeit sofort durch Diffusion abgebaut werden. Wenn diese Überlegungen richtig sind, muß sich aus dem Vergleich von stationär bzw. dynamisch gemessenen Werten für Auszehrung und Aufsättigung annähernd die gleiche zusätzliche Sauerstoffmenge ergeben.

Berechnung aus der Auszehrungsphase

Der tatsächliche momentane O_2-Bedarf der Kultur entspricht der Größe $r_{O_2\,stat}$. Während der Auszehrung gilt

$$r_{O_2\,dyn} = \frac{dc_L}{d\tau} \quad (1)$$

Wenn $r_{O_2\,dyn} < r_{O_2\,stat}$, muß der Zusammenhang lauten

$$r_{O_2\,stat} = r_{O_2\,dyn} + \frac{dc_z}{d\tau} \quad (2)$$

Die Zeitdauer des Auszehrungsvorganges beträgt

$$\tau_{Aus} = \frac{c_{L\infty}}{r_{O_2\,dyn}} \quad (3)$$

so daß man über

$$c_z = \int_0^{\tau_{Aus}} (r_{O_2\,stat} - r_{O_2\,dyn})\, d\tau \quad (4)$$

wegen $r_{O_2\,stat} - r_{O_2\,dyn} = \text{const.}$ erhält

$$c_z = c_{L\infty}\left(\frac{r_{O_2\,stat}}{r_{O_2\,dyn}} - 1\right) \quad (5)$$

Berechnung aus der Aufsättigungsphase

Es gilt der Zusammenhang

$$\frac{dc_L}{d\tau} = k_La(c^*_{O_2} - c_L) - r_{O_2} \quad (6)$$

oder

$$c_L = -\frac{1}{k_La}\left(\frac{dc_L}{d\tau} + r_{O_2}\right) + c^*_{O_2} \quad (7)$$

der in Abb. 5 graphisch dargestellt ist.

Für $k_La_{stat} \neq k_La_{dyn}$ kann man schreiben

$$c_L = c^*_{O_2} - \frac{1}{k_La_{dyn}}\left(\frac{dc_L}{d\tau} + r_{O_2\,dyn}\right) \quad (8)$$

$$c_L = c^*_{O_2} - \frac{1}{k_La_{stat}}\left(\frac{dc_L}{d\tau} + \frac{dc_z}{d\tau} + r_{O_2\,stat}\right) \quad (9)$$

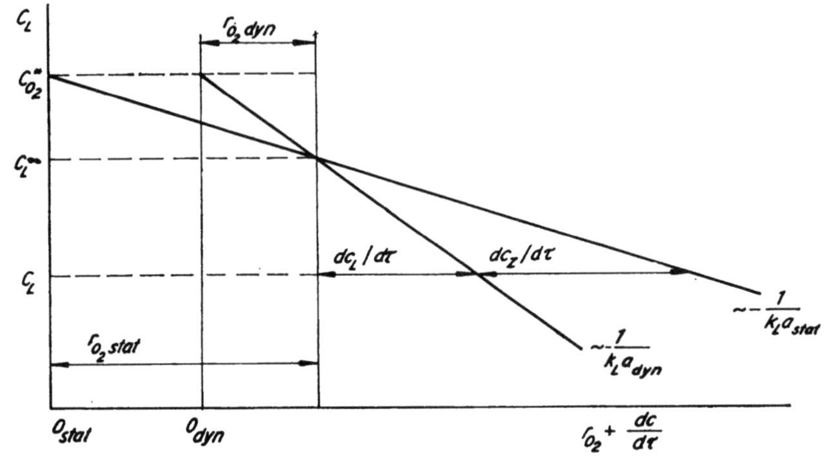

Abb. 5. Graphische Darstellung von Gleichung (7)

(8) und (9) werden gleichgesetzt und man erhält

$$\frac{dc_z}{d\tau} = \frac{dc_L}{d\tau}\left(\frac{k_L a_{\text{stat}}}{k_L a_{\text{dyn}}} - 1\right) + r_{O_2\,\text{dyn}} \cdot \frac{k_L a_{\text{stat}}}{k_L a_{\text{dyn}}} - r_{O_2\,\text{stat}} \tag{10}$$

Da

$$r_{O_2\,\text{stat}} = r_{O_2\,\text{dyn}} \cdot \frac{k_L a_{\text{stat}}}{k_L a_{\text{dyn}}} \tag{11}$$

ergibt sich

$$\frac{dc_z}{d\tau} = \frac{dc_L}{d\tau}\left(\frac{k_L a_{\text{stat}}}{k_L a_{\text{dyn}}} - 1\right) \tag{12}$$

und damit

$$c_z = \int_0^{\tau_{\text{Auf}}} \frac{dc_L}{d\tau}\left(\frac{k_L a_{\text{stat}}}{k_L a_{\text{dyn}}} - 1\right) \cdot d\tau \tag{13}$$

Gleichung (13) wird ausgehend von der Aufsättigungskurve graphisch integriert. c_z ist auf das Fermentationsvolumen bezogen.

Bestimmung von Absolutwerten für c_z

Die Auswertung der Versuche nach den beschriebenen Verfahren ergibt folgendes Bild. Für den Lysinversuch errechnet man einen Mittelwert für

Auszehrung $c_z = 9{,}95 \cdot 10^{-3}\ \text{kgO}_2/\text{m}^3$

Aufsättigung $c_z = 11{,}6 \cdot 10^{-3}\ \text{kgO}_2/\text{m}^3$

ohne signifikante Abhängigkeit der Streuung von anderen Versuchsparametern. Das bedeutet, daß bei einer Sättigungskonzentration von $c^*_{O,E} = 7{,}75 \cdot 10^{-3}\ \text{kgO}_2/\text{m}^3$ zusätzlich die 1,3 bis 1,5fache Menge Sauerstoff freigesetzt bzw. angelagert wird.

Bezogen auf das Biomassevolumen beträgt die Zusatzkonzentration 0,34 bzw. $0{,}41 \frac{kgO_2}{m^3_{Zelle}}$.

In [20] ist für die Hefe „*Saccharomyces cerevisiae*" ein Wert von 0,0286 kg O_2/m^3_{Zelle} angegeben.

Für die Turimycinversuche zeigt sich dagegen eine deutliche Abhängigkeit von der stationären Gelöst-O_2-Konzentration $c_{L\infty}$ (Abb. 6). Bei unterschiedlichen Absolutwerten ergibt sich der gleiche Anstieg über $c_{L\infty}$. Die Unterschiede resultieren möglicherweise aus verschiedenen Turimycingehalten. Die in Abb. 6 eingezeichnete Kurve entspricht einer Verdoppelung der Auszehrungs- bzw. Aufsättigungszeit.

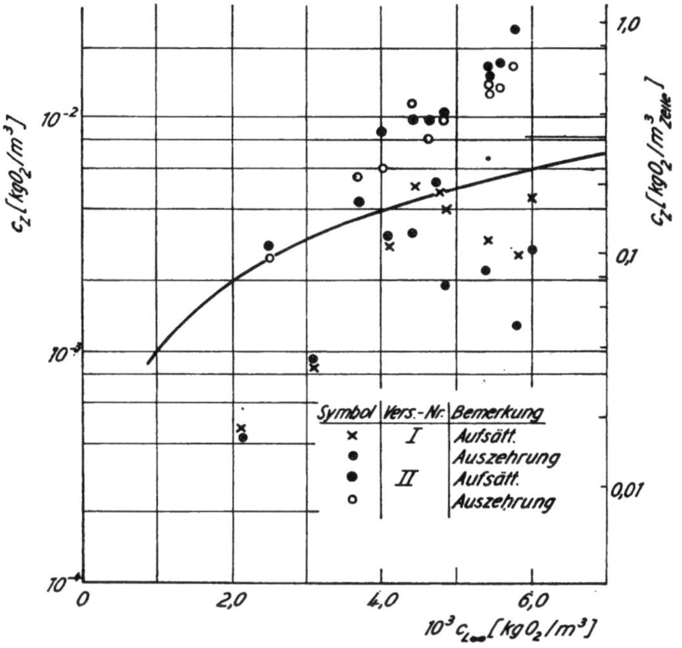

Abb. 6. Zusätzlich freigesetzter bzw. gelöster Sauerstoff bei den dynamischen Messungen während der Turimycinfermentationen

Sowohl bei Lysin als auch bei Turimycin werden im Rahmen der Streubreite der Messung aus der Auszehrungs- und Aufsättigungsphase die gleichen Werte für den zusätzlich freigesetzten bzw. gelösten Sauerstoff errechnet.

Schlußbemerkungen

Die beschriebenen Ergebnisse zeigen, daß die dynamische Methode zur Messung von k_La-Werten in Fermentationssystemen nur bedingt einsetzbar ist. Das beobachtete Phänomen einer zum in der flüssigen Phase gelösten Sauerstoff zusätzlichen Sauerstoffanlagerung ist meßtechnisch eindeutig nachweisbar, kann jedoch vorerst biologisch oder biochemisch nicht erklärt werden. Dazu und zur genaueren Bestimmung der Abhängigkeit von der Fermentationszeit sind weitere Untersuchungen erforderlich.

Nomenklatur

c_z	kgO_2m^{-3}	Zusätzlicher Gelöstsauerstoff
c	—	Relative Konzentration
c_L	kgO_2m^{-3}	Gelöstsauerstoffkonzentration
$c^*_{O_2}$	kgO_2m^{-3}	Sättigungskonzentration
H	$kgO_2m^{-3}Pa^{-1}$	Henrysche Konstante
k_La	h^{-1}	Volumetrischer Stoffübergangskoeffizient
K_G	—	Korrekturfaktor Gasverweilzeit
K_S	—	Korrekturfaktor Sondenzeitkonstante
P_R	W	Rührerleistung
$\bar{p}$	Pa	Mittlerer Druck
r_{O_2}	$kgO_2m^{-3}h^{-1}$	Sauerstoffübergangsrate
V_G	m^3	Gasvolumen in der Dispersion
$\dot{V}_G$	m^3h^{-1}	Belüftungsrate
V_L	m^3	Flüssigkeitsvolumen
ψ_{O_2,CO_2}	—	Partialdruck
φ_G	—	Hold-up
τ	s	Zeit
ϱ_{O_2}	kgm^{-3}	Dichte

Indizes

E	Eintrittszustand
N	Normzustand
Auf	Aufsättigung
Aus	Auszehrung
stat	stationär
dyn	dynamisch
∞	Endzustand

Den Mitarbeitern des ZIMET Jena, besonders Herrn Dr. Forberg, sei für die Unterstützung und Beratung während der Versuche und Auswertung gedankt.

Eingegangen: 13. 12. 1982

Literatur

[1] Vant't Riet, K.: Ind. Eng. Chem. Process Des. Dev. 18 (1979) 3, 357.
[2] Käppeli, O., Fiechter, A.: Biotechnol. Bioeng. 22 (1980), 1509.
[3] Linek, V., Vacek, V.: Biotechnol. Bioeng. 18 (1976), 1537.
[4] Dang, N. D. P. u. a.: Biotechnol. Bioeng. 19 (1977), 853.
[5] Linek, V., Benes, P.: Biotechnol. Bioeng. 20 (1978), 903.
[6] Hassan, J. T. M., Robinson, C. W.: Biotechnol. Bioeng. 19 (1977), 661.
[7] Votruba, J., Sobota, M.: Biotechnol. Bioeng. 18 (1976), 1815.
[8] Votruba, J. u. a.: Biotechnol. Bioeng. 20 (1978), 913.
[9] Votruba, J. u. a.: Biotechnol. Bioeng. 19 (1977), 435.
[10] Heijnen, J. J. u. a.: Biotechnol. Bioeng. 22 (1980), 1945.
[11] Linek, V.: Biotechnol. Bioeng. 19 (1977), 983.
[12] Dunn, J. J., Einsele, A.: J. Appl. Chem. Biotechnol. 25 (1975), 707.
[13] Bandyopadhyay, B. u. a.: Biotechnol. Bioeng. 9 (1967), 533.
[14] Paca, J., Gregr, V.: J. Appl. Chem. Biotechnol. 27 (1977), 155.
[15] Hsu, K. H. u. a.: Biotechnol. Bioeng. 17 (1975), 499.

[16] LIN, C. H. u. a.: Biotechnol. Bioeng. 18 (1976) 1557.
[17] KNÖLL, H. u. a.: Verfahren zur Gewinnung eines Makrolid-Antibiotikums, DDR-Patent 84450 vom 5. 3. 1969.
[18] Verfahren zur Herstellung von Lysin mit Hilfe von Mikroorganismen, WP DD 140056.
[19] LIEPE, F., MATSCHKE, J.: Experimentelle Methoden zur Bestimmung der Stoffübergangsintensität in Gas-in-Flüssig-Dispersionen, Interner Bericht IH Köthen 1981.
[20] NIKOLAJEV, P., SOKOLOV, D.: Teorija i praktika nepereryvnovo kultiwirovanija mikroorganismov, Izdatelstvo „Nauka", Moskau 1980, S. 189.
[21] LIEPE, F., MEUSEL, W.: Zu einigen Fragen der Sauerstoffkinetik bei Fermentationsprozessen in flüssiger Phase, Zwischenbericht IH Köthen, 31. 1. 1982.

Acta Biotechnol. 4 (1984) 3, 246

Book Review

J. Coombes

International Biotechnology Directory 1984

London: Macmillan Publishers Ltd., 1983, 426 S., 40 Ł

Wie ein Untertitel verspricht, enthält das Directory Angaben über Organisationen, Gesellschaften, Forschungsrichtungen und Produkte. Mit ihm wird ein Nachschlagewerk mit hohem Informationsgehalt vorgelegt, das eine wertvolle spezifische Ergänzung zur umfangreichen Literatur zur Biotechnologie darstellt und dem auf diesem Gebiet Arbeitenden Sucharbeit erspart und neue Kenntnisse vermittelt. Es dient gleichzeitig durch die Aufführung der zahlreichen bestehenden Organisationen der internationalen Zusammenarbeit und erleichtert neu in der Biotechnologie Tätigen den Kontakt zu diesen.
Im Directory werden systematisierte Hinweise gegeben sowohl zur klassischen Biotechnologie, wie zur Antibiotika- und Vitaminproduktion, als auch zur modernen Biotechnologie, d. h. zur Gentechnologie, Energiegewinnung und Biomasseproduktion.
Im Teil I des Directory werden internationale Organisationen und Informationsmöglichkeiten (Zeitschriften; Datenbasen) aufgeführt. Teil II enthält nationale Profile (Gesellschaften, Industriebetriebe, Forschungsinstitute, Universitäten, Handelsvereinigungen). Im Teil III werden Produkte, Ausrüstungen und Forschungsgebiete beschrieben. Dazu zählen u. a. Alkohole, Enzyme, Futtermittel, Pharmazeutika, Biochemikalien; Fermentoren, Pumpen, Meßgeräte, EDV-Software; Kultursammlungen; Zellkultivierung, Screening, Genetik, Fermentation, Ingenieurwesen, Analytik.
Das Directory ist damit ein Wegweiser nicht nur für die auf dem Gebiet der Biotechnologie Arbeitenden, sondern auch für die deren Ergebnisse Nutzenden, z. B. in der Landwirtschaft, in der Medizin, beim Umweltschutz und in der pharmazeutischen und chemischen Industrie.
Im Directory 1984 sind Angaben zu Westeuropa, Nordamerika und Japan enthalten. Die vorgesehene jährliche Aktualisierung ist sehr zu begrüßen. Sein Wert würde zweifellos noch erhöht, wenn die Liste der internationalen und nationalen Organisationen, die unvollständig ist, ergänzt würde und die wichtigen Aktivitäten solcher Länder wie UdSSR, DDR und CSSR aufgenommen würden. Auch sollte die Liste der Zeitschriften um die in diesen Ländern zum Thema Biotechnologie erscheinenden ergänzt werden. Es kann dem Wunsche Ausdruck verliehen werden, daß dem guten Directory 1984 ein noch besseres folgt.

L. Dimter

Acta Biotechnol. 4 (1984) 3, 247–254

Zur Maßstabsübertragung von aeroben mikrobiologischen Prozessen
Teil II. Wärmetransport

F. Kudrewizki

Institut für Technische Mikrobiologie
DDR-1017 Berlin, Alt-Stralau 62

Summary

An universal consideration is proposed for calculation of the heat exchange of industrial fermenters at which the energy-intake is usual as main parameter in the formulas. On this way mathematical models developed, useful for heat transport during mixing by means of mixer with and without aeration, during boiling and for bubble column, are declared. With these models it is possible to describe Non-Newtonian fluids, which rheological conditions in the concrete apparatus is formed by the intaken energy. Stable rubbers are proposed for treatment of high viscous fluids and a half-empirical model is offered for its calculation.

Bei der Entwicklung neuer Bioreaktoren oder bei der Anpassung vorhandener Apparaturen an neue Prozesse ist in der Regel die Frage des Wärmetransportes zu lösen. Sowohl die intensive Wärmezufuhr bei der thermischen Sterilisation des Mediums als auch die Abführung der bei der mikrobiellen Reaktion entstandenen Wärme bedingen eine richtige Auslegung der Wärmetransportflächen und der Parameter des Wärmeträgers. Die Fragen des Wärmetransportes in homogenen Newton'schen Medien in Rührkesseln sind in der Literatur vielfach beschrieben [1—5]. Dagegen bestehen bei den Kenntnissen über den Wärmetransport in den Systemem Gas—Flüssigkeit und besonders in den Nicht-Newton'schen Medien noch einige Lücken. Die auf diesem Gebiet vorhandenen Verallgemeinerungen [6, 7] haben hauptsächlich nur einen empirischen Charakter. Aus diesem Grunde wird eine rechnerische Analyse des Wärmetransportes in Gas—Flüssigkeit-Systemen unter den noch nicht untersuchten Bedingungen im wesentlichen erschwert.
In früheren Arbeiten [8—10] wurde gezeigt, daß es unter Berücksichtigung der Energiedissipation des Gases gelingt, halbempirische Gleichungen des Wärmetransportes und der effektiven Zähigkeit der Nicht-Newton'schen Medien aufzustellen. Es erscheint zweckmäßig, auch bei der Analyse des Wärmetransportes in den Fermentoren ähnlich heranzugehen.

Effektive Zähigkeit der Nicht-Newton'schen Medien beim Mischen

In vielen praktischen Fällen werden Fermentationsmedien angewendet, die dem rheologischen Ostwald-Gesetz untergeordnet sind:

$$\tau = K_\eta \left(\frac{dU}{dy}\right)^m \tag{1}$$

In Form einer Abhängigkeit der effektiven Zähigkeit von der Scherspannung wird dieses Gesetz durch die Gleichung (2) ausgedrückt:

$$\eta_{\text{eff}} = K_\eta^{1/m} \tau^{1-\frac{1}{m}} \tag{2}$$

Die Scherspannungen entstehen beim Mischen als Ergebnis der Energiedissipation. Die zähe Energiedissipation wird mit der Gleichung (3) [11] erfaßt:

$$\varepsilon = \eta \left(\frac{dU}{dy}\right)^2, \tag{3}$$

die auch in der integralen Form (4)

$$\varepsilon = A\nu\varrho \left(\frac{U}{l}\right)^2 \tag{4}$$

aufgeschrieben werden kann.
Durch Verknüpfung und Umstellung der Gleichungen (2) und (4) wird die Abhängigkeit (5) erhalten:

$$\eta_{\text{eff}} = K_\eta^{\frac{2}{m+1}} \left(\frac{\varepsilon}{A}\right)^{\frac{m-1}{m+1}} \tag{5}$$

Die Anwendung der Gleichung (5) ist günstig, wenn der Makromaßstab der Turbulenz durch die Apparatekonstruktion prädestiniert wird und von den Arbeitsbedingungen unabhängig bleibt. Dieser Fall tritt z. B. in Rührkesseln ein, bei denen der mittlere Wert der Energiedissipation durch die Rührerleistung

$$\varepsilon = \frac{N}{V} = \frac{1}{V} K_N \cdot \varrho \cdot n^3 \cdot d_2^5 \tag{6}$$

bestimmt wird. Unter Berücksichtigung dieser Gleichung nimmt der Ausdruck für die effektive Zähigkeit folgendes Aussehen an:

$$\eta_{\text{eff}} = K_\eta^{\frac{2}{m+1}} \left(\frac{K_N \varrho n^3 d_2^5}{V \cdot A}\right)^{\frac{m-1}{m+1}} \tag{7}$$

Durch die Ergebnisse von Begačev et al. [12] ist es in letzter Zeit möglich geworden, den Leistungsbeiwert K_N praktisch für alle Rührertypen, die in Nicht-Newton'schen Medien arbeiten, zu bestimmen und damit die effektive Zähigkeit nach der Formel (7) zu berechnen. Das schließt aber nicht die Anwendbarkeit der traditionellen Methode aus, die auf der Ausnutzung der Abhängigkeiten (8) [13] basiert.

$$K_N = f(Re_m) \qquad \text{und} \qquad \left(\frac{dU}{dy}\right) = A_1 n \tag{8}$$

Für die annähernde Analyse der Blasensäulen mit großem Durchmesser kann man die effektive Zähigkeit unter Berücksichtigung der Abhängigkeit (9)

$$\tau = \varrho U^2 = \varrho(\overline{W}' g\, d_b)^{2/3} \tag{9}$$

abschätzen, wobei der Pulsationsmaßstab durch den Blasendurchmesser ersetzt werden kann. Wird (9) mit (2) verknüpft, entsteht der Koeffizient der effektiven Zähigkeit beim Begasen (10)

$$\eta_{\text{eff}} = [g(W' g\, d_b)^{2/3}]^{1-\frac{1}{n}} \cdot K_\eta^{1/n} \tag{10}$$

Die effektive Zähigkeit von Nicht-NEWTON'schen Gas-Flüssigkeits-Strömungen beim Fließen in engen vertikalen Kanälen (Röhren) kann man auch durch die Verknüpfung der Abhängigkeit (5) mit der Energiedissipation der Gas-Flüssigkeits-Strömungen bestimmen (11):

$$\varepsilon = W'(1 - \varphi)\,\varrho g, \tag{11}$$

wobei (12) entsteht:

$$\eta_{\text{eff}} = K_\eta^{\frac{2}{m+1}} \left[\overline{W}' g (1 - \varphi) \frac{1}{A}\right]^{\frac{m-1}{m+1}} \tag{12}$$

Der Ausdruck für φ ist aus dem ersten Teil dieser Arbeit [14] zu entnehmen.

Intensivierung des Wärmetransportes durch Energieeintrag

In Analogie zu früheren Arbeiten [8, 9, 14] wird die Analyse des Wärmetransportes auf der Grundlage der Energiedissipation in bezug auf das Wärmetransportmodell aufgebaut. Es wird davon ausgegangen, daß bei der ungeordneten Bewegung ein System der lokalisierten konvektiven Strömungen der willkürlichen Maßstabslänge l existiert, die den Wärmetransport beim An- und Umströmen der Wand intensivieren. Wird der Strömungsmaßstab so gewählt, daß die Bewegung als laminar bezeichnet werden kann, entsteht folgende Abhängigkeit für die lokale Größe des Wärmetransportkoeffizienten (13):

$$\frac{\alpha l}{\lambda} = C \left(\frac{U \cdot l}{\nu}\right)^{0,5} \mathrm{Pr}^{1/3} \tag{13}$$

Bei der senkrechten Anströmung ist $C = 1{,}13$, bei der Strömung entlang der Wand ist $C = 0{,}33$ [17]. Wird von der gleichen Wahrscheinlichkeit bei der räumlichen Bewegung ausgegangen, erhält man die mittlere Größe $C = 0{,}5$. Wenn die Ausdrücke für U und l aus (4) abgeleitet werden, entsteht die Gleichung (14) für den Wärmeübergang von der Wand:

$$\frac{\alpha l}{\lambda} = 0{,}5 \left[\left(\frac{\varepsilon}{\nu \varrho A}\right)^{1/2} \frac{l^2}{\nu}\right]^{1/2} \mathrm{Pr}^{1/3} \tag{14}$$

Aus dieser Gleichung ist ersichtlich, daß die Auswahl des Maßstabes keine Rolle spielt und der A-Wert auch nur einen sehr geringen Einfluß auf die Größe des Wärmeabgabekoeffizienten hat. Eine günstige einheitliche Form für Gleichung (14) zur Anwendung für NEWTON'sche und Nicht-NEWTON'sche Medien wird durch die Gleichung (15) dargestellt:

$$\frac{\alpha}{\lambda} \left(\frac{a^2}{g}\right)^{1/3} = 0{,}23 \left[\frac{\varepsilon}{\varrho g (g \cdot \nu_{\text{eff}})^{1/3}}\right]^{1/4} \left(\frac{a}{\nu_{\text{eff}}}\right)^{1/3} \left(\frac{\eta}{\eta_w}\right)^{0,14} \tag{15}$$

Das Verhältnis $\left(\frac{\eta}{\eta_w}\right)^{0,14}$ ist für die Berücksichtigung der Richtung des Wärmestromes eingeführt.

Wärmeübergang von der Wand in Blasensäulen

Der Ausdruck der Energiedissipation hat in Blasensäulen mit großem Durchmesser folgendes Aussehen (16):

$$\varepsilon = \overline{W}' \varrho g \tag{16}$$

Damit nimmt Gleichung (15) die endgültige Form (17) an:

$$\frac{\alpha}{\lambda}\left(\frac{a^2}{g}\right)^{1/3} = 0{,}23\left[\frac{\overline{W}'}{(\nu_{\text{eff}}\cdot g)^{1/3}}\right]^{1/4}\left(\frac{a}{\nu_{\text{eff}}}\right)^{1/3}\left(\frac{\eta}{\eta_w}\right)^{0,14} \tag{17}$$

Wärmeübergang von der Wand in die Gas-Flüssigkeits-Medien in engen Kanälen

In diesem Fall wird die Energie der Gas-Flüssigkeits-Strömung mit dem Ausdruck (11) beschrieben. Im Unterschied zum Begasen in Blasensäulen wird der Wert für die Konstante C in der Gleichung (13) ebenso wie bei der Strömung entlang des Eingangsabschnittes der Platte oder des Rohres angenommen [8, 9]. Man erhält zuerst (18):

$$\frac{\alpha\delta}{\lambda} = 0{,}33\left[\delta^2(1-\varphi)\left(\frac{\overline{W}'g}{A\nu}\right)^{1/2}\right]^{1/2}\mathrm{Pr}^{1/3} \tag{18}$$

und anschließend nach der einfachen Umwandlung und dem Vergleich mit den Experimentaldaten [8, 9] (19):

$$\frac{\alpha}{\lambda}\left(\frac{a^2}{g}\right) = 0{,}155\left[\frac{\overline{W}'}{(g\cdot\nu_{\text{eff}})^{1/3}}\right]^{1/4}(1-\varphi)^{1/2}\left(\frac{a}{\nu_{\text{eff}}}\right)^{1/3}\left(\frac{\eta}{\eta_w}\right)^{0,14} \tag{19}$$

Wärmeübergang von der Wand in Rührkesseln bei $d_2 < D$

Der Wärmeübergang in Rührkesseln wird auch mit Gleichung (15) beschrieben, die nach dem Verknüpfen mit Formel (6) folgende Form (20) annimmt:

$$\frac{\alpha}{\lambda}\left(\frac{a^2}{g}\right)^{1/3} = 0{,}23\left[\frac{K_N\cdot n^3\,d_2^{\,5}}{g(g\cdot\nu_{\text{eff}})^{1/3}}\right]^{1/4}\left(\frac{a}{\nu_{\text{eff}}}\right)^{1/3}\left(\frac{\eta}{\eta_w}\right)^{0,14} \tag{20}$$

Diese Gleichung ist für verschiedene Rührertypen mit den Verhältnissen $\frac{d_2}{D} < 1$ besonders günstig. Sie ist aber nicht geeignet für die Berechnungen der Apparate mit Schabrührern $\left(\frac{d_2}{D} = 1\right)$, die infolge hoher Wärmeübergangsintensität als Wärmeaustauscher für zähe und schlecht transportable Medien eingesetzt werden können.

Wärmeübergang in Apparaten mit Schabrührer ($d_2 = D_1$)

Von verschiedenen Autoren [18—20] wurde für die Berechnungen des Wärmeüberganges von einer Wand, an der ein Schaber entlanggeführt wird, die Abhängigkeit

$$\mathrm{Nu} = CB^{0,5}\cdot\mathrm{Pe}_Z^{\,0,5} \tag{21}$$

vorgeschlagen. Diese Beziehung wurde aus der Lösung der Fourier-Gleichung auf der Grundlage des Modells der instationären Wärmeleitfähigkeit im halbabgegrenzten Raum abgeleitet. Dabei wurde angenommen, daß die Flüssigkeitstemperatur hinter dem Schaber der Temperatur im Strömungskern gleich ist. In [18—20] sind aber solche Schaber untersucht worden, die durch die hydrodynamische Widerstandskraft des Mediums an die Wand angedrückt worden sind. Bei solcher Konstruktion erfolgt bei der Minderung der Reinolds-Zahl $\mathrm{Re} < 300$ eine bemerkbare Reduzierung der Intensität des Wärmetransportes infolge des Entstehens eines unregelbaren Spaltes zwischen dem Schaber und der Wand sowie der Verschlechterung des Wärmetransportes im Strömungskern. Der

Einfluß des zweiten Faktors ist von der Schaberkonstruktion unabhängig und kann in großen Apparaten durch den Einbau zusätzlicher Rührer beseitigt werden. Im Gegensatz dazu hängt der Spalt zwischen Wand und Rührer mit der Rührerkonstruktion zusammen und kann durch die Anwendung von Federn [5, 9], insbesondere bei Schabern mit dem Winkel < 90° zur Bewegungsrichtung, der in diesem Fall als Wischer bezeichnet werden soll, beseitigt oder wesentlich vermindert werden. Dabei entsteht ein kleiner Spalt bzw. eine feine Flüssigkeitsschicht zwischen dem Wischer und der Wand, der eine Abnutzung sowohl des Wischers als auch der Wand, hervorgerufen durch die gegenseitige Reibung, stark reduziert.
Zur Beschreibung des Wärmeüberganges von der Wand wurden unter Berücksichtigung des thermischen Widerstandes der Schicht und nach dem Vergleichen mit den Experimentaldaten die Berechnungsgleichungen (22) und (23) erhalten [5, 9]:

Für $$2000 \leqq G_\beta \leqq \frac{d_2 \sin\beta \cdot \mathrm{tg}\,\beta \cdot K_f}{4\,\mathrm{w} \cdot \eta_{\mathrm{eff}}} \leqq 200000$$
$$\mathrm{Nu} = 0{,}087 \cdot \mathrm{Pe}_z^{0{,}5} \cdot G_\beta^{0{,}19} \tag{22}$$

Für $$450 \leqq G_\beta \leqq 2000$$
$$\mathrm{Nu} = 0{,}0035 \cdot \mathrm{Pe}_z^{0{,}5} \cdot G_\beta^{0{,}6} \tag{23}$$

Die effektive Zähigkeit wird auch hier, wie in [20] beschrieben, mit der Gleichung (24) berechnet, die nach der METZNER-Methode ermittelt worden war:

$$\eta_{\mathrm{eff}} = K_\eta (110 B^{0{,}5} \cdot n)^{m-1} \tag{24}$$

Wärmeübergang beim Kochen

Bei der Mediensterilisation in Apparaten ohne Rührung, z. B. in Blasensäulen, kann ein Kochprozeß an den beheizten Flächen entstehen. Es ist in diesem Zusammenhang von Interesse, eine Analyse des Wärmeüberganges durchzuführen. Die Untersuchungen, die mit den Modellmedien Polyethylen- und CMC-Lösungen, $\eta = 5 - 6000$ Pa · s, durchgeführt worden sind [10], haben gezeigt, daß der Wärmeübergangskoeffizient nur in geringem Maße vom Wärmestrom und von der Zähigkeit abhängt:

$$\alpha \sim f(\eta^{-0{,}18}, q^{n_1}), \tag{25}$$

wobei $n_1 < 0{,}2$ ist. Dabei konnte kaum $q = 1200$ W/m² erreicht werden, obwohl Δt bis $\Delta t = 40$ °C hochgezogen wurde. Es ist festzustellen, daß der Wärmetransport durch die konzentrierte Schicht des gelösten Feststoffanteils, die das Blasensieden im Wandfilm unterdrückt, an der Wand limitiert wird. Andererseits wird der Wärmetransport durch kontinuierliches Auswaschen des Filmes und durch die Konvektion infolge des Dampfblasenaufstieges im Kern des Stromes stimuliert. Berechnungen des Wärmeüberganges kann man mit den Formeln (17) oder (19) durchführen, wobei η_w als Zähigkeit der für die angegebenen Druckverhältnisse maximal konzentrierten Lösungen angenommen wird und die lineare Gasgeschwindigkeit durch den Wärmestrom bestimmt wird.
In der Praxis ist das Kochen an der Wand zu vermeiden, weil damit keine Intensivierung des Wärmetransportes zu erwarten ist. Eine wesentliche Intensivierung des Wärmetransportes kann dagegen erreicht werden, wenn einer Wandanlagerung der konzentrierten Lösung vorgebeugt wird. Das kann durch die Erhöhung des Druckes erreicht wer-

den, wobei die Kochtemperatur des Mediums erhöht und der Wärmetransport durch das eingeleitete Gas intensiviert wird. Der Wärmeübergang ist dabei nach Formel (17) oder (19) zu berechnen.

Wärmeübergang beim Begasen und Kochen in einem Rührkessel

Eine Intensivierung des Wärmeüberganges von der Wand an die zähen Gas-Flüssigkeits-Gemische und auch an die zähen kochenden Medien ist in Rührkesseln effektiv realisierbar [9, 10]. Die Berechnungsformeln werden aus der Addition der Energiedissipation abgeleitet [9, 21]:

$$\sum \varepsilon = \varepsilon_R + \varepsilon' + \varepsilon_K \tag{26}$$

Daraus ergibt sich für den Wärmeübergangskoeffizient die Formel (27):

$$\alpha_A = \sqrt[4]{\alpha_R{}^4 + \alpha'^4 + \alpha_K{}^4} \tag{27}$$

Eine kleine Abweichung von dieser Formel entsteht gesetzmäßig beim Mischen Nicht-NEWTON'scher Medien mit den Wischrührern, da die Energiedissipation als einziges bestimmendes Kriterium dabei kaum angewendet werden kann. Außerdem wird die effektive Zähigkeit beim Mischen verringert. Die Untersuchungen haben gezeigt, daß, obwohl die Voraussetzungen für die Anwendung der Formel gegeben sind, die Potenzen etwas kleiner sind. In Abb. 1 sind die Versuchsergebnisse mit den CMC-Lösungen in einem Apparat mit einem Wischrührer dargestellt, die mit der Formel (28) befriedigend erfaßt werden:

$$\alpha = \sqrt[3]{\alpha_R{}^3 + \alpha'^3} \tag{28}$$

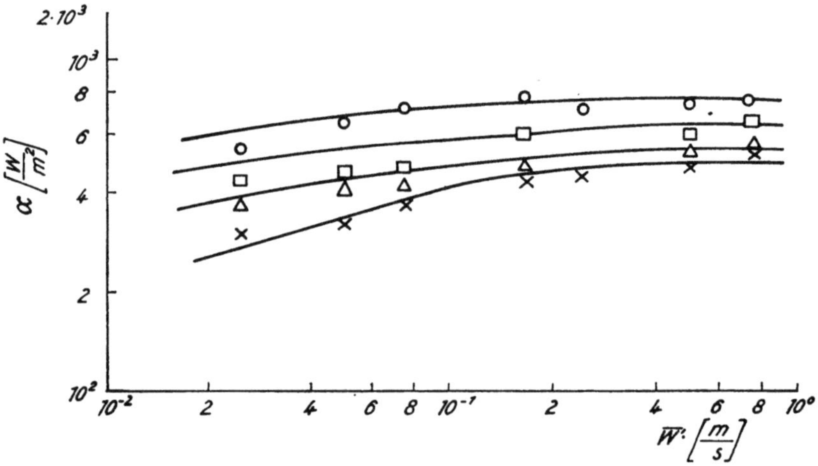

Abb. 1. Wärmeübergang von der Wand zu der 11%igen wäßrigen CMC-Lösung im begasten Rührkessel mit Schabrührer Korr.: anstelle $\alpha\left[\frac{W}{m^2}\right]$ lies $\alpha\left[\frac{W}{m^2 \cdot \text{grad}}\right]$

Bezeichnung	n, s⁻¹	
×	0,0	Punkte — Experimentaldaten ($K_\eta = 1{,}85$, $m = 0{,}7$)
△	0,167	Kurve — Berechnungen nach den Formeln
□	0,466	(12, 19, 22—24)
○	2,2	

Das oben angegebene Vorgehen wird bei der Auswertung von Versuchsdaten in verschiedenen Fällen bei erzwungener Konvektion oft verwendet [22].
Wie schon angedeutet, ist als eine wichtige Aufgabe bei der Auslegung der industriellen Fermentoren die Abfuhr der Reaktionswärme zu realisieren, d. h. eine richtige Auswahl der Wärmeaustauschflächen und der Parameter des Wärmeträgers zu treffen. Dafür können die oben angegebenen Formeln direkt genutzt werden.

Nomenklatur

Formelzeichen	Erklärung	Einheit
a	Temperaturleitzahl	m^2/s
A, A_1, C	Konstanten	—
B	Wiescherzahl	—
D, d_2, d_b	Durchmesser des Apparates, des Rührers, der Blasen	m
g	Erdbeschleunigung	m/s^2
H	Flüssigkeitshöhe	m
K_η	OSTWALD-Konstante	Pa · s
K_N	Leistungsbeiwert des Rührers	—
l	Pulsationsmaßstab	m
m	Fließindex	—
n	Drehzahl des Rührers	s^{-1}
U	Geschwindigkeit der erzwungenen Konvektion	m/s
$\frac{dU}{dy}$	Schergefälle	s
V	Apparateinhalt	m^3
W'	lineare Geschwindigkeit	m/s
α	Wärmeübergangskoeffizient	$kJ/m^2 \cdot s \cdot grad$
δ	Kanalbreite	m
ε	spezifischer Energieeintrag	kW/m^3
η	dynamische Viskosität	Pa · s
λ	Wärmeleitkoeffizient	kJ/m · s · grad
ν, ν_{eff}	kinematische bzw. effektive kinematische Zähigkeit	m^2/s
ϱ	Dichte	kg/m^3
τ	Schubspannung	Pa

Eingegangen: 14. 12. 1983

Literatur

[1] LIEPE, F., WEISSGÄRBER, H.: Stoffvereinigen in fluiden Phasen. In: Verfahrenstechnische Berechnungsmethoden, Teil 4/2. Leipzig: VEB Deutscher Verlag für Grundstoffindustrie, 1978.
[2] POGGEMANN, R. u. a.: Chem. Ing. Tech. 10 (1979) 10, 948.
[3] BEGAČEV, V. I., BRAGINSKIJ, L. N., PAVLUŠENKO, I. S.: Gidrodinamika i massoobmen v chimičeskoj apparature, Wissenschaftliche Abhandlung über Maschinenbau, Leningrad 1967, S. 56—66.
[4] BRAGINSKIJ, L. N., BEGAČEV, V. I.: Teoretičeskie osnovy chimičeskoj technologii T III. 1969, Nr. 2, S. 245.
[5] KONSETOV, V. V., KUDREWIZKI, F. M., NOVICHKOV, A. N.: Chimičeskoe mašinostroenie, Wissenschaftliche Abhandlung des Forschungsinstituts für Chemiemaschinenbau, 1972, 67.
[6] LIEPE, F., MÖCKEL, H.-O.: Stoffvereinigen in fluiden Phasen. In: Verfahrenstechnische Berechnungsmethoden, Teil 4/1. Leipzig: VEB Deutscher Verlag für Grundstoffindustrie, 1978.

[7] STEIFF, A., WEINSPACH, P. M.: Chem. Ing.-Tech. **49** (1977) 2, 157. Synopse 446—477.
[8] KONSETOV, V. V., KUDREWIZKI, F. M.: Inženerno-fizičeskij zurnal (1972) 4, 667.
[9] KUDREWIZKI, F. M., KONSETOV, V. V.: Žurnal prikladnoj chimii (1972) 6, 1269.
[10] KONSETOV, V. V., KUDREWIZKI, F. M., NOVICHKOV, A. N.: Promyšlennost' sintetičeskogo kaučuka (1970) 5, 7.
[11] HINZE, I. O.: Turbulence — An Introduction to its mechanism and theory, Mc Graw Hill Book Company, Inc., New York: 1959.
[12] BEGAČEV, V. I., GURVIČ, A. R., BRAGINSKIJ, L. N.: Teoretičeskie osnovy chimičeskoj technologii, Theoretische Grundlagen der chemischen Technik, **14** (1980) Nr. 1.
[13] METZNER, A. B., OTTO, R. E.: AICHE J. **1** (1961) 3.
[14] KUDREWIZKI, F.: Acta Biotechnol. **4** (1984) 89.
[15] JANOWSKII, E. A.: Promotionsarbeit A, Leningrader Lensowjetinstitut, 1970.
[16] KUDREWIZKI, F.: Chem. Techn. (1982) Nr. 5.
[17] SCHLICHTING, R.: Teorija pograničnogo sloja (Grenzschichttheorie), Inostrannaja literatura, Moskau, 1956.
[18] COOL, I.: Trans. Inst. Chem. (London) **36** (1958), 253.
[19] HARRIOT, P.: Chem. Eng. Prog. Symp. Ser. **55** (1959) 29, 137.
[20] BRAGINSKIJ, L. N., BEGAČEV, V. I., PAVLUŠENKO, I. S.: Žurnal prikladnoi chimii **37** (1964) 1984.
[21] DOMANSKIJ, I. V., SOKOLOV, V. N.: Teoretičeskie osnovy chimičeskoj technologii (1968) Nr. 5.
[22] KUTATELADSE, S. S.: Teoretičeskie osnovy teploobmena, Maschiz, Moskau-Leningrad, 1962.
[23] KUTATELADSE, S. S., Borišanskij, V. M.: Spravočnik po teploobmenu, Goschimizdat, 1959.

Acta Biotechnol. 4 (1984) 3, 255—266

Исследование физиологических характеристик и технологических показателей роста Candida utilis на среде с этанолом

С. Б. Мутафов[1], В. К. Ерошин[2], И. Г. Минкевич[2]

[1] Институт микробиологии БАН, ул. Академика Георгия Бончева, бл. 26, 1113 София, НРБ
[2] Институт биохимии и физиологии микроорганизмов АН СССР, 142292 г. Пущино, СССР

Summary

The growth of the yeast *Candida utilis* VKM-Y-2332 was investigated during the cultivation on a mineral medium with addition of ethanol in the regime of the chemostat and the pH-auxostat.
The composition of the macroelements of the biomass produced by *Candida utilis* VKM-Y-2332 was studied.
Stoichiometric equations of the growth process of the examined yeast strain on ethanol were presented.
The growth of *Candida utilis* on ethanol and glucose was also characterized by the way of comparision.

Биомасса дрожжей, выращенных на этаноле, в настоящее время рассматривается как перспективный компонент рациона человека [10]. В качестве возможных продуцентов пищевых дрожжей на этаноле предполагаются штаммы *Candida utilis*, [20, 21]. Поэтому целью настояшего исследования является определение ряда физиологических констант роста дрожжей *C. utilis*, а также оценка эффекта использования этанола в качестве субстрата на продуктивность ферментационного оборудования.

Экспериментальная часть

Методика

Штамм *C. utilis* ВКМ-Y-2332 поддерживали в пробирках на сусло-агаре. В качестве посевного материала использовали односуточную культуру дрожжей, выращенных на той же среде в матрасах при температуре 37°C.
Исследование проводили в режимах хемостата и pH-ауксостата. При культивировании в режиме pH-ауксостата питательная среда одновременно является титрантом и подается по сигналу разбаланса между pH уставки и pH культуральной жидкости [8, 16]: pH подаваемой среды обязательно выше, чем pH уставки. В установившемся режиме удельная скорость роста μ равна скорости разбавления D, однако устойчивость равенства поддерживается автоматом по регулировке pH, а не клетками, как в хемостате. При этом все компоненты питательной среды присутствуют в культуральной жидкости в избытке. Концентрация биомассы x определяется тем количеством аммиака, которое должно быть потреблено из подаваемой среды, чтобы понизить ее pH до значения pH уставки. Следовательно, x зависит от буферной емкости среды.

Концентрация остаточного органического субстрата S в ферментере равна $S = S_0 - x/Y_s$, где Y_s — выход клеток из субстрата по массе, S_0 — концентрация субстрата в подаваемой среде. Изменяя величину S_0, можно получить разные значения S и исследовать влияние этой концентрации на удельную скорость роста клеток при высоком содержании субстрата в культуральной жидкости. В этом отношении pH-ауксостат подобен турбидостату, но выгодно отличается от последнего устойчивостью своей работы: скорость разбавления в pH-ауксостате весьма стабильна во времени.

Учитывая сказанное, мы определяли в pH-ауксостате константу ингибирования роста K_i и максимальную удельную скорость роста μ_m. В хемостате по зависимости μ от S находили константу насыщения удельной скорости роста субстратом K_s и также μ_m (используя другую совокупность экспериментальных данных); по зависимости Y_s от μ — значения максимального выхода по субстрату Y_s^m и коэффициента поддержания клеток m_s (см. ниже).

Культивирование проводили на ферментационной установке АК-3 с рабочим объемом 1,5 л, оборудованной механическим пеногасителем и системой для автоматической коррекции pH. Последняя использовалась в pH-ауксостате для регулирования потока питательной среды. Во всех режимах культивирования поддерживали температуру 36°, pH = 5,0 и аэрацию воздухом 1 л/мин при оборотах мешалки 800 об/мин. В режимах хемостатного культивирования пользовались минеральной питательной средой следующего состава (в мг/л): KH_2PO_4 — 1000, $MgSO_4 \cdot 7H_2O$ — 500, NH_4Cl — 800, $FeSO_4 \cdot 7H_2O$ — 16, $CaCl_2 \cdot 6H_2O$ — 0,66, $ZnSO_4 \cdot 6H_2O$ — 5, $CuSO_4 \cdot 5H_2O$ — 0,16, $MnSO_4 \cdot 5H_2O$ — 0,15, $CoCl_2 \cdot 6H_2O$ — 0,18 и C_2H_5OH — 5000; pH = 4,5. Заданное значение pH культуры поддерживали 5%-ным раствором аммиака. Скорость протока контролировали прямым измерением бюретой скорости расхода питательной среды.

В режиме pH- ауксостата использовалась та же солевая среда, pH которой был повышен до 9 путем добавления аммиака. Во избежание образования осадков компоненты среды были разделены: фосфат и аммиак в одном растворе, прочие соли — в другом. Подача среды осуществлялась по раздельным шлангам двухканального перистальтического насоса.

Концентрацию этанола в культуральной жидкости и в питательной среде определяли на хроматографе Chrom-4 (KOVO, ЧССР) с пламенно-ионизационным детектором (сорбент Porapac QS) [1]. Количество абсолютно сухой биомассы (АСБ) определяли фильтрованием на мембранных фильтрах № 9 (Сынпор, Хемапол, ЧССР), высушиванием и взвешиванием клеток, содержащихся в 5 мл культуры. Поскольку учет испарившегося этанола затруднителен, величину выхода по субстрату Y_S в опытах на pH-ауксостате не определяли. Макроэлементный состав дрожжевой биомассы определяли в клетках, выращенных в хемостате при скорости разбавления $D = 0{,}25$ час$^{-1}$. Количество золы определялось сожжением биомассы в муфельной печи, а содержание углерода, водорода и азота — автоматическим C, H, N — анализатором (KOVO, ЧССР). Биомасса до анализа сохранялась в лиофилизированном виде. Анализы выполнены в лаборатории и их статистическая оценка сделана по методу парного линейного регрессионного анализа [3]. Значение выхода биомассы по субстрату рассчитывали по формуле:

$$Y_S = x/(S_0 - S) \tag{1}$$

Результаты и их обсуждение

Опытные данные о культивировании *C. utilis* ВКМ-Y-2332 в режимах хемостата и pH-ауксостата представлены в табл. 1. Эти данные были использованы для получения физиологических констант культуры.

Из табл. 1 видно, что в условиях хемостатного культивирования концентрация остаточного субстрата в культуральной среде S возрастает с увеличением объемной скорости разбавления $D = \mu$. Это возрастание описывается формулой Моно [18]:

$$\mu = \mu_m S/(K_S + S) \tag{2}$$

Представление экспериментальных данных в системе координат $1/S$, $1/\mu$ (рис. 1-A) и результаты линейного регрессионного анализа зависимости обратной величины удельной скорости роста μ от обратной величины концентрации субстрата S показали следующие значения величин константы насыщения и максимальной удельной скорости роста:

$K_s = (2{,}5 \pm 0{,}6) \cdot 10^{-3}$ г/л и $\mu_m = 0{,}545 \pm 0{,}031$ час$^{-1}$. Коэффициент корреляции независимой переменной и функции отклика составлял при этом 0,87, что свидетельствует об адекватности функции (2) полученным экспериментальным данным.

Культивирование в режиме pH-ауксостата показало, что при высоких значениях концентрации этанола в культуральной среде S удельная скорость роста культуры μ снижается, и при концентрациях этанола между 19 и 24 г/л ее значение равно приблизительно половине значения максимальной удельной скорости роста, найденной в опытах в хемостате. Из литературы известно, что этанол является типичным неконкурентным ингибитором роста [7, 11]. В случаях неконкурентного ингибирования роста константа ингибирования K_i отвечает той концентрации ингибитора, при которой удельная скорость роста клеток равна половине ее максимального значения. В рассматриваемом случае это означает, что ожидаемое значение величины константы ингибирования K_i находится в пределах 19—24 г/л. Точное значение величины константы ингибирования K_i получили с использованием математической модели неконкурентного ингибирования роста субстратом [7]:

$$\mu = \mu_m S K_i / (K_s K_i + S K_i + S^2) \tag{3}$$

Так как значение константы насыщения K_s, полученное в опытах в хемостате, пренебрежимо мало по сравнению с ожидаемым значением величины константы ингибирования K_i, $K_s \ll K_i$, можно упростить уравнение (3) для достаточно больших концентраций субстрата, то есть для $S > 0{,}5$ г/л:

$$\mu = \mu_m K_i / (K_i + S) \tag{4}$$

Последнее уравнение преобразуется к линейному виду с коэффициентами регрессии $1/\mu_m$ и $1/\mu_m K_i$:

$$\frac{1}{\mu} = \frac{1}{\mu_m} + \frac{1}{\mu_m K_i} \cdot S \tag{5}$$

Представляя экспериментальные данные в систему координат S, $\frac{1}{\mu}$ (рис. 1-Б), получили следующие значения величин максимальной удельной скорости роста и константы ингибирования (коэффициент корреляции 0,82):

$\mu_m = 0{,}543 \pm 0{,}025$ час$^{-1}$ и $K_i = 20{,}2 \pm 2{,}3$ г/л.

Вопрос о субстратном ингибировании роста дрожжей этанолом слабо изучен. По данным Aiba с соавт. [11], значение константы ингибирования роста хлебопекарных дрожжей этанолом равно примерно 23 г/л. Эта величина близка к найденной нами. Значения максимальной удельной скорости роста μ_m, полученные нами в опытах в хемостате и в pH-ауксостате, практически одинаковы. Далее мы будем пользоваться величиной $\mu_m = 0{,}544$ час$^{-1}$. Надо отметить, что это значение максимальной удельной скорости роста μ_m невозможно получить ни в каком реальном режиме роста, так как ингибирование этанолом начинает сказываться еще до достижения полного насыщения величины μ по этому же субстрату. Самое высокое значение удельной скорости роста $\mu_k = 0{,}532$ час$^{-1}$ может быть достигнуто при критической концентрации субстрата $S_k = \sqrt{K_s K_i} = 0{,}225$ г/л. Дальнейшее увеличение концентрации этанола в культуральной среде приводит к уменьшению удельной скорости роста. Это хорошо видно из рис. 1-В, где зависимость удельной скорости роста

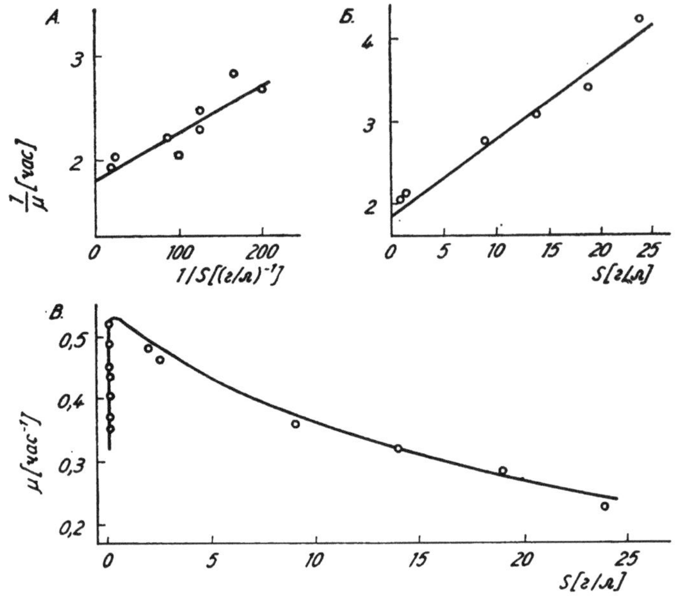

Рис.. 1 Влияние концентрации субстрата (S) на удельную скорость роста (μ) *C. utilis* ВКМ-Y-2332.

А. Влияние низких концентраций этанола на удельную скорость роста. Экспериментальные данные представлены в системе обратных координат согласно уравнению (2) (см. текст).

Б. Влияние высоких концентраций этанола на удельную скорость роста. Экспериментальные данные представлены согласно упрощенной модели неконкурентного ингибирования роста субстратами (уравнение (4), см. текст).

В. Влияние широкого диапазона концентраций этанола на удельную скорость роста. Кривая зависимости $\mu(S)$ построена согласно модели неконкурентного ингибирования роста субстратом (уравнение (3), см. текст) с использованием найденных в опыте значений констант $K_s = 2{,}5$ мг/л, $K_i = 20{,}2$ г/л и $\mu_m = 0{,}544$ час$^{-1}$.

дрожжей *C. utilis* ВКМ-Y-2332 от концентрации этанола представлена в обычной системе координат S, μ согласно функции (3) с использованием найденных в эксперименте значений физиологических констант роста μ_m, K_s и K_i. Экспериментальные данные хорошо укладываются на кривую, что свидетельствует об адекватности использованной модели неконкурентного ингибирования роста субстратом.

Из представленных в табл. 1 экспериментальных данных видно, что с уменьшением удельной скорости роста дрожжей μ значения выхода биомассы по субстрату снижаются. Это снижение выхода по субстрату Y_s при низких значениях удельной скорости роста объясняется относительным увеличением доли субстрата, идущей на поддержание жизнедеятельности клеток, и описывается известной моделью влияния расходов на поддержание на выходы по субстрату в зависимости от удельной скорости роста, данной Иерусалимским [2] и Пертом [19].

$$\frac{1}{Y_s} = \frac{1}{Y_s^m} + \frac{m_S}{\mu} \qquad (6)$$

Табл. 1. Концентрация остаточного субстрата (S) и выход сухой биомассы (Y_s) дрожжей *C. utilis* ВКМ-Y-2332 из этанола при различных значениях удельной скорости роста (μ)

хемостат	μ час$^{-1}$	0,030	0,052	0,060	0,084	0,100	0,133	0,180	0,261	0,284
	S г/л	а	а	а	а	а	а	а	а	а
	Y_s г/г	0,525	0,591	0,605	0,635	0,640	0,665	0,674	0,687	0,693
	μ час$^{-1}$	0,329	0,352	0,371	0,403	0,433	0,450	0,485	0,491	0,517
	S г/л	а	0,006	0,005	0,008	0,008	0,012	0,010	0,049	0,054
	Y_s г/г	0,694	0,711	0,687	0,722	0,694	0,680	0,690	0,714	0,693
pH-ауксостат	μ, час$^{-1}$	0,483	0,469	0,364	0,322	0,289	0,235			
	S, г/л	2	2,5	9	14	19	24			
	Y_s, г/г	б	б	б	б	б	б			

а) Концентрация этанола в культуральной среде ниже предела чувствительности хроматографа.

б) Не определялось.

где Y_s^m — максимальный выход по субстрату, m_s — коэффициент поддержания клеток (расход субстрата на поддержание, г на 1 г АСБ в час).

Представление экспериментальных данных в системе координат $1/\mu$, $1/S$ (рис. 2-А) и их линейный регрессионный анализ дали следующие значения максимального выхода по субстрату и коэффициента поддержания: $Y_s^m = 0{,}716 \pm 0{,}003$ и $m_s = 0{,}0153 \pm 0{,}0006$ час$^{-1}$ при значении коэффициента корреляции 0,67. На рис. 2-Б представлена зависимость выхода дрожжей *C. utilis* ВКМ-Y-2332 от удельной скорости роста μ в обычной системе координат μ, Y_s. Кривая зависимости $Y_s(\mu)$ построена согласно функции (6) с использованием найденных в опыте значений максимального выхода по субстрату Y_s^m и коэффициента поддержания m_s. Равномерное расположение экспериментальных значений выхода по субстрату Y_s и значение коэффициента корреляции $0{,}67 > 0{,}5$, свидетельствуют о соответствии функции (6) экспериментальным данным. Из рис. 2-Б видно также, что при достаточно высоких значениях удельной скорости роста μ значения выхода по субстрату Y_s составляют примерно 0,7. Это значение выхода дрожжей *C. utilis* при культивировании на среде с этанолом чаще всего встречается в литературе [9, 15].

Следует отметить, что расходы субстрата на поддержание жизнедеятельности клеток имеют энергетическую природу. В то же время модель Иерусалимского-Перта описывает эти расходы с помощью коэффициента m_s, измеряемого в единицах массы. Энергетическое описание эффективности роста и влияние на нее процессов поддержания клеток разработано Минкевичем и Ерошиным [6, 17] и Эриксоном с соавт. [12, 13] на основе специальной единицы измерения, учитывающей одновременно и массу, и энергосодержание органических веществ. Первоначальный вариант этой единицы, называвшийся доступным электроном (ДЭ: available electron, AE) был позднее усовершенствован [4] и в обобщенном виде называется редоксоном (RO) (для C, H, O, N — соединений число RO равно числу ДЭ). Согласно теории материально-энергетического баланса [6, 12, 13, 17] энергетический выход биомассы из суб-

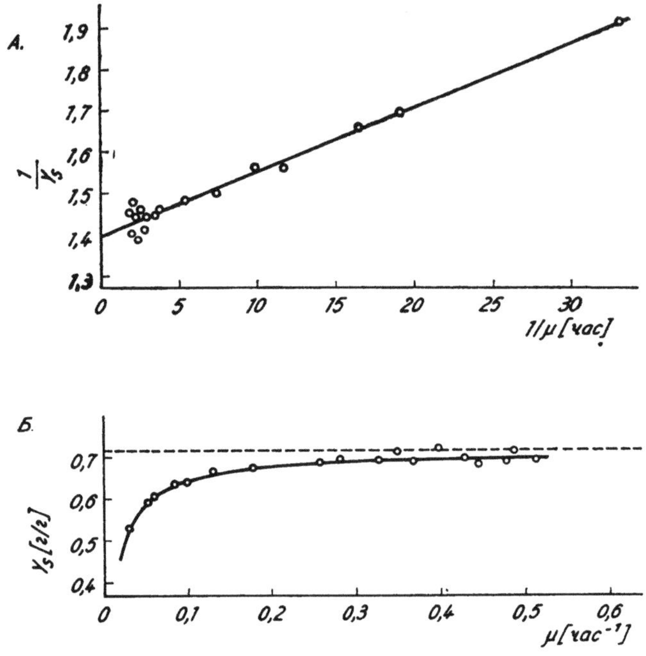

Рис. 2. Влияние удельной скорости роста (μ) на выход биомассы по субстрату (Y_s).

А. Экспериментальные данные представлены в системе обратных координат согласно уравнению Иерусалимского-Перта (6) (см. текст).

Б. Экспериментальные данные представлены в обычной системе координат. Кривая зависимости $Y_s(\mu)$ построена согласно уравнению (11) (см. текст) с использованием найденных в опыте значений физиологических констант $Y_s^m = 0{,}716$ и $m_s = 0{,}0153$ час$^{-1}$.

страта η равен отношению числа редоксонов синтезированной биомассы к числу редоксонов потребленного субстрата. Величина η связана с выходом клеток по массе $\boldsymbol{Y_s}$:

$$\eta = \frac{\sigma_b \gamma_b}{\sigma_s \gamma_s} \cdot Y_s \tag{7}$$

где σ_b и σ_s — весовая доля углерода соответственно в биомассе и в субстрате, γ_b и γ_s — степень восстановленности этих веществ. Степень восстановленности биомассы и субстрата γ зависит от макроэлементного состава этих соединений и вычисляется из нормализованных по углероду брутто-формул веществ (формула вещества, представленная для 1 грамм-атома углерода, $CH_pO_nN_q$):

$$\gamma = 4 + p - 2n - 3q \tag{8}$$

Удельная скорость роста влияет на энергетический выход согласно формуле, аналогичной (6) [13]:

$$\frac{1}{\eta} = \frac{1}{\eta^m} + \frac{m_e}{\mu} \tag{9}$$

Энергетический коэффициент поддержания m_e (число эквивалентов RO субстрата, затраченных в час на поддержание биомассы, содержащей 1 эквивалент RO) связан с m_s:

$$m_e = \frac{\sigma_s \gamma_s}{\sigma_b \gamma_b} \cdot m_s \tag{10}$$

Максимальный энергетический выход биомассы η^m связан с Y_s^m формулой идентичной формуле (7).

Таблица 2 объединяет данные анализа макроэлементного состава биомассы дрожжей *C. utilis* ВКМ-Y-2332, выращенных на минеральной питательной среде с этанолом, и коэффициенты расчета энергетической эффективности роста опытного штамма. Пересчитывая Y_s^m и m_s с помощью уравнений (7) и (10), получаем:

$$\eta^m = 0{,}432 \pm 0{,}001 \quad \text{и} \quad m_e = 0{,}025 \pm 0{,}001 \text{ час}^{-1}$$

Результаты исследования роста дрожжей *C. utilis* ВКМ-Y-2332 на среде с этанолом показывают, что величина константы субстратного ингибирования K_i многократно превышает константы насыщения K_s, $K_i \gg K_s$. В условиях хемостатного культивирования, когда субстрат потребляется почти полностью, его концентрация несоизмеримо меньше значения коэффициента ингибирования, $S \ll K_i$, и этанол почти не проявляет ингибирующего влияния на скорость роста. Поэтому реальное наибольшее значение скорости роста $\mu_k = 0{,}528$ час$^{-1}$ близко к $\mu_m = 0{,}544$ час$^{-1}$. Величина константы насыщения $K_s = 2$—3 мг/л на 3 порядка (примерно в 1000 раз) меньше концентрации этанола в подаваемой среде S_0. Поэтому остаточным содержанием этанола в культуральной жидкости при лимитировании этим субстратом можно пренебречь и считать, что клетки полностью потребляют этанол.

Зависимость выхода по субстрату от удельной скорости роста, согласно (6), имеет вид:

$$Y_s = Y_s^m \mu / (\mu + m_s Y_s^m) \tag{11}$$

При культивировании данного штамма на среде с этанолом $m_s Y_s^m \sim 0{,}01$ час$^{-1}$, тогда как практический интерес представляют скорости роста порядка 0,5 час$^{-1}$.

Табл. 2. Макроэлементный состав биомассы дрожжей *C. utilis* ВКМ-Y-2332, выращенной на среде с этанолом и коэффициенты расчета материально-энергетического баланса роста

C, %	H, %	N, %	Зола, %	O, %[a]
44,62	6,80	7,85	6,67	34,06

Нормализованная по углероду брутто-формула биомассы	Весовая доля углерода в биомассе σ_b	Степень восстановленности биомассы, γ_b	Коэффициент перевода из массовых в материально-энергетические единицы описания баланса роста дрожжей на среде с этанолом, $\sigma_b\gamma_b/\sigma_s\gamma_s$
$CH_{1,829}O_{0,573}N_{0,15}$	0,446	4,231	
Нормализованная по углероду брутто-формула этанола	Весовая доля углерода в этаноле, σ_s	Степень восстановленности этанола, γ_s	0,603
$CH_3O_{0,5}$	0,521	6	

a) Вычислено по разности между 100% и содержанием других компонентов.

При таких значениях μ, согласно (11) выход биомассы Y_s весьма близок к своему максимальному значению Y_s^m (см. рис. 2-Б).

На рис. 3 приведены зависимости концентрации биомассы x, остаточного субстрата S и производительности ферментера $P = Dx$ для хемостатных режимов, то есть при скоростях разбавления от $D = 0$ до $D = \mu_k$. Зависимости рассчитаны на основе уравнений для $\mu(S)$ (3) и $Y_s(\mu)$ (11), которые, как сказано выше, хорошо удовлетворяют экспериментальным данным. Использовались вышенайденные значения физиоло-

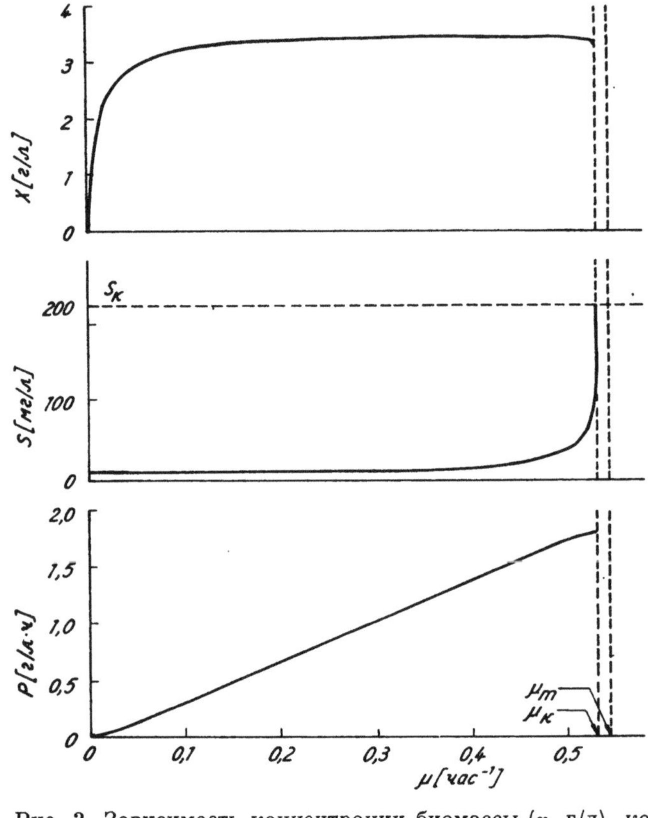

Рис. 3. Зависимость концентрации биомассы (x, г/л), концентрации остаточного субстрата (S, мг/л) и продуктивности хемостата ($P = Dx$, г/л. час) при различных значениях удельной скорости роста (μ). (Расчет на основе уравнений (3) и (11) с использованием найденных в опыте значений физиологических констант $\mu_m = 0{,}544$ час$^{-1}$, $K_s = 2{,}5$ мг/л, $K_i = 20{,}2$ г/л, $Y_s^m = 0{,}716$ и $m_s = 0{,}153$ час$^{-1}$.

гических констант штамма μ_m, K_s, K_i, Y_s^m и m_s. Концентрация остаточного субстрата S найдена из (3) при условии $\mu = D$, концентрация биомассы x — из соотношения $x = (S_0 - S)Y_s$. Снижение x при малых скоростях разбавления обусловлено падением Y_s из-за влияния затрат на поддержание. Как видно из рис. 3, при $\mu >$ 0,2 час$^{-1}$ биомасса практически не меняется и равна своему максимальному значению. Причиной этого и для x, и для Y_s является малость величины m_s (см. выше). Падение x при больших скоростях разбавления, а также резкое возрастание концентрации остаточного субстрата происходят в узком диапазоне D вблизи значения скорости разбавления $D = \mu_m$. Узость этого диапазона обусловлена тем, что

$K_s \ll S_0$. По той же причине максимальное значение производительности хемостата P достигается при скорости разбавления D также весьма близкой к μ_k.

Отдельно следует отметить, что энергетическая эффективность роста данного штамма η близка к 0,4, то есть при росте на среде с этанолом клетки содержат в своем органическом веществе примерно 40% энергии потребленного субстрата. Для сравнения укажем, что часто достигаемое при росте на глюкозе и других углеводах значение $Y_s = 0{,}5$ соответствует $\eta \sim 0{,}6$, то есть в 1,5 раза выше, чем на этаноле. Причины этого различия и возможность повышения η на этаноле в данной статье не обсуждаются. Однако, учитывая, что и этанол и углеводы представляют интерес как сырье для крупномасштабного производства биомассы, проведем сравнение ряда технологических параметров выращивания дрожжей *C. utilis* на этих субстратах, используя найденные выше характеристики опытного штамма. Для такого сравнения примем весовой выход на этаноле равным $Y_s^{э} = 0{,}7$; а на глюкозе или других углеводах равным $Y_s^{г} = 0{,}5$ [14, 15]. Примем также следующие вполне реальные допущения: а) константа насыщения K_s на глюкозе, и соответственно концентрация остаточной глюкозы, как и на этаноле, много меньше, чем концентрация субстрата в подаваемой среде S_0; б) макроэлементный состав биомассы дрожжей *C. utilis*, выращенных на среде с глюкозой, не отличаются существенно от соответствующего состава биомассы, выращенной на среде с этанолом [14]; в) рост на среде с глюкозой в лимите по этому субстрату не связан с накоплением в культуральной жидкости стехиометрически значимых количеств органических продуктов. При культивировании дрожжей *C. utilis* на среде с этанолом в хемостате нами показано, что уксусная кислота, ацетальдегид, этилацетат (то есть наиболее вероятные побочные продукты роста) отсутствовали в культуральной жидкости в хроматографически детектируемых количествах.

Находя выход клеток по углероду y_c по формуле [17],

$$y_c = \frac{\sigma_b}{\sigma_s} Y_s \tag{12}$$

получаем следующие уравнения баланса роста [17]. На среде с этанолом:

$$\begin{aligned} &CH_3O_{0,5} + 0{,}090\,NH_3 + 0{,}866\,O_2 \\ &= 0{,}599\,CH_{1,829}O_{0,573}N_{0,151} + 0{,}401\,CO_2 + 1{,}087\,H_2O \end{aligned} \tag{13}$$

На среде с глюкозой:

$$\begin{aligned} &CH_2O + 0{,}084\,NH_3 + 0{,}410\,O_2 \\ &= 0{,}558\,CH_{1,829}O_{0,573}N_{0,151} + 0{,}442\,CO_2 + 0{,}616\,H_2O \end{aligned} \tag{14}$$

Отметим, что в уравнениях (13) и (14) учтены лишь те химические элементы (C, H, O, N), которые составляют основу баланса роста. Поэтому получить из уравнений (13) и (14) те значения Y_s, которые были использованы для вычисления y_c по формуле (12), можно, если дополнительно учесть содержание золы в биомасс.

Уравнения (13) и (14) позволяют вычислить выход биомассы по кислороду Y_0 (г/г). Обозначим через b стехиометрический коэффициент при O_2 (0,866 в (13) и 0,410 в (14)), M — молекулярный вес субстрата, c — число атомов углерода в его молекуле. Количество субстрата, содержащее 1 грамм-атом углерода, равно M/c грамм. При его потреблении синтезируется $\frac{M}{c} \cdot Y_s$ г биомассы и потребляется, согласно (13) и (14), $32b$ г кислорода. Отсюда

$$Y_0 = Y_s M/32bc \tag{15}$$

Подставляя $Y_s^э = 0{,}7$ и $Y_s^г = 0{,}5$, получаем $Y_0^э = 0{,}581$ и $Y_0^г = 1{,}143$. Соответственно, расход кислорода на 1 г выросшей биомассы на этаноле и на глюкозе равен $1/Y_0^э = 1{,}721$ и $1/Y_0^г = 0{,}875$. Отсюда видно, что расход кислорода при росте на этаноле в расчете на одно и то же количество полученной биомассы вдвое (точнее, в $Y_0^г/Y_0^э = 1{,}97$ раза) выше, чем на глюкозе. При аэробном росте теплопродукция клеток пропорциональна потребленному кислороду [17]. Поэтому и количество физиологического тепла в расчете на полученную биомассу при культивировании на среде с этанолом будет вдвое выше, чем при культивировании на среде с глюкозой.

Метод материально-энергетического баланса роста микроорганизмов дает простой способ выражения теплопродукции клеток при аэробном их выращивании [17]. Поскольку редоксоны (в старых терминах доступные электроны) субстрата и биомассы имеют практически одинаковый энергетический уровень $Q_0 = 113$ кДж на эквивалент RO, то теплопродукция обусловлена лишь движением редоксонов на кислород. Количество кислорода, потребленного в расчете на 1 г полученной биомассы, равно $1/Y_0$ г, то есть $\frac{4}{32} \cdot \frac{1}{Y_0} = 1/8\, Y_0$ г-эквивалентов. Тогда теплопродукция на 1 г выращенной биомассы равна

$$Q_0 \cdot \frac{1}{8Y_0} = 14{,}13 \cdot \frac{1}{Y_0} \text{ кДж/г АСБ} \tag{16}$$

Это составляет для этанола 24,3 кДж/г АСБ, а для глюкозы 12,4 кДж/г АСБ.

Отметим, что теплопродукция клеток при аэробном росте, рассчитанная не на 1 г АСБ, а на 1 г потребленного кислорода, постоянна и не зависит ни от эффективности роста, ни от других параметров процесса. Она равна 113 кДж/г-экв. O_2 или около 1,23 кДж/г O_2 [5]. Поэтому, замена субстрата (например, глюкозы на этанол) при условии, что рост лимитирован кислородом, а скорость массопередачи не изменилась, (одни и те же температура, давление, обороты мешалки и т.д.) не изменяет общую скорость теплопродукции в единице объема ферментера.

Поскольку величины K_s при культивировании на среде с этанолом и на среде с глюкозой малы, производительность ферментера в лимите по этим субстратам равна [5]:

$$P_s = DS_0Y_s \tag{17}$$

Эта величина возрастает при увеличении концентрации органического субстрата S_0 до тех пор, пока не произойдет переключение лимита с этого субстрата на кислород. Тогда производительность начинает определяться следующим соотношением [5]:

$$P_0 = K_LaC^*Y_0 \tag{18}$$

где K_La — коэффициент массопереноса кислорода, C^* — растворимость кислорода в среде, K_LaC^* — максимальная скорость растворения кислорода. Величина P_0 является значением производительности ферментера, максимальным по концентрации подаваемого субстрата. Для одного и того же ферментера, в одних и тех же массообменных условиях, согласно (18), отношение значений максимальной производительности при культивировании на средах с этанолом и с глюкозой равно

$$\frac{P_0^э}{P_0^г} = \frac{Y_0^э}{Y_0^г} \tag{19}$$

что для вышеприведенных значений эффективности роста равно приблизительно 0,5.

Наиболее рациональным является культивирование в режиме, соответствующем переходу от лимита по органическому субстрату к лимиту по кислороду. Тогда достигается одновременное полное использование и подаваемого со средой субстрата, и массообменной мощности ферментера. В таком режиме $P_s = P_0$. Используя (17) и (18), находим значение абсолютной скорости подачи органического субстрата (на единицу рабочего объема ферментера), соответствующее границе между двумя лимитами:

$$DS_0 = \frac{K_L a C^* Y_0}{Y_s} \tag{20}$$

Отношение этих величин для этанола и глюкозы равно:

$$\frac{(DS_0)^{э}}{(DS_0)^{г}} = \frac{Y_0^{э} Y_s^{г}}{Y_0^{г} Y_s^{э}} \tag{21}$$

что составляет 0,36 для вышеприведенных значений эффективностей роста.

Из-за более низкого энергетического выхода биомассы *C. utilis* на этаноле производительность ферментационной установки на этом субстрате вдвое ниже, чем на глюкозе. Такая установка, работая в экономичном режиме (на грани двух лимитирований), способна переработать в единицу времени количество этанола, равное 36% (по массе) от количества перерабатываемого в таком же режиме углеводного сырья.

Выводы, сделанные из настоящего исследования, имеют прикладное значение при планировании лабораторных экспериментов. Но, мы надеемся, они могут оказаться полезными и при разработке процесса выращивания дрожжей *C. utilis* на этаноле с целью получения биомассы пищевого назначения на ферментационных установках с известными технологическими показателями.

Использованные обозначения

Индексы **г** и **э** в тексте означают, что намеченная ими величина относится соответственно к культивированию на среде с глюкозой или на среде с этанолом.

- b — стехиометрический коэффициент потребления кислорода в молях на 1 г-атом С использованного субстрата, безразмерно.
- c — число атомов С в молекуле использованного субстрата, безразмерно.
- C^* — растворимость кислорода в культуральной среде, г.л$^{-1}$.
- D — объемная скорость разбавления, час$^{-1}$.
- K_i — константа ингибирования роста субстратом, г.л$^{-1}$.
- $K_L a$ — коэффициент массопереноса кислорода, час$^{-1}$.
- K_s — константа насыщения скорости роста субстратом, г.л$^{-1}$.
- M — молекулярная масса органического субстрата, г
- m_e — удельная скорость расхода энергии на поддержание, час$^{-1}$ [уравнение (9) и (10)].
- m_s — удельная скорость расхода субстрата на поддержание, час$^{-1}$ [уравнение (6) и (11)].
- n — число г-атомов кислорода на 1 г-атом С биомассы или субстрата, безразмерно.
- p — число г-атомов водорода на 1 г-атом С биомассы или субстрата, безразмерно.
- P_s — продуктивность ферментера в лимите по органическому субстрату, г · л$^{-1}$ · час$^{-1}$ [уравнение (17)].
- P_0 — продуктивность ферментера в лимите по кислороду, г.л$^{-1}$ · час$^{-1}$ [уравнение (18)].
- q — число г-атомов азота на 1 г-атом С биомассы, безразмерно.
- Q_0 — теплообразование на 1 эквивалент кислорода [$Q_0 = 113$ кДж (экв.)]
- S — концентрация органического субстрата в культуральной среде, г.л$^{-1}$.
- S_k — критическая концентрация этанола $\left(S_k = \sqrt{K_i K_s}\right)$, г.л$^{-1}$
- S_0' — концентрация органического субстрата в питательной среде, г.л$^{-1}$.
- x — концентрация биомассы в культуральной среде, г.л$^{-1}$.
- y_c — выход биомассы по углероду, безразмерно.

Y_0 — выход биомассы в расчете на массу потребленного кислорода, безразмерно.
Y_s — выход биомассы в расчете на массу потребленного органического субстрата, безразмерно.
Y_s^m — максимальный выход биомассы в отсутствие затрат на поддержание, безразмерно [уравнение (6) и (11)].
γ — степень восстановленности углерода в органическом субстрате или в биомассе, безразмерно [уравнение (8)].
η — энергетический выход биомассы, безразмерно, [уравнение (7)].
η^m — максимальный энергетический выход биомассы, безразмерно [уравнение (9)].
μ — удельная скорость роста, час$^{-1}$.
μ_k — значение удельной скорости роста, реально достигаемое при критической концентрации субстрата S_k, час$^{-1}$.
μ_m — максимальная удельная скорость роста, час$^{-1}$.
σ_b — доля массы углерода в биомассе, безразмерно.
σ_s — доля массы углерода в органическом субстрате, безразмерно.

Eingegangen: 4. 3. 1983

Литература

[1] Джеффери П., Киппинг П.: Анализ газов методом газовой хроматографии. М., Мир, 1976.
[2] Иерусалимский Н. Д.: Основы физиологии микробов. М., Изд-во АН СССР, 1963.
[3] Львовский Е. Н.: Статистические методы построения эмпирических формул. М., Высшая школа, 1982.
[4] Минкевич И. Г.: — В сб.: Лимитирование и ингибирование микробиологических процессов. НЦБИ АН СССР, Пущино, 1980.
[5] Минкевич И. Г., Ерошин В. К.: Микробиологическая пром-сть, (1974) 7 (115), 1.
[6] Минкевич И. Г., Ерошин В. К.: Успехи совр. биол. **82** (1976) 1 (4), 103.
[7] Перт С. Дж.: Основы культивирования микроорганизмов и клеток. М., Мир, 1978.
[8] Печуркин Н. С., Позмогова И. Н. и Терсков И. А.: Прикл. биохим. и микробиол. **5** (1969) 158.
[9] Позмогова И. Н., Минкевич И. Г., Уткина И. С., Лирова С. А.: Изв. АН СССР, сер. биол. (1979) 6, 858.
[10] Получение белка пищевого достоинства на основе новых сырьевых источников. Обзорная информация. Глав. управл. микробиол. пром-сти Сов. Мин. СССР, М, 1979.
[11] Aiba, S., Shoda, M., Nagatani, M.: Biotechnol. Bioeng. **10** (1968) 845.
[12] Erickson, L. E., Minkevich, I. G., Eroshin, V. K.: Biotechnol. Bioeng. **20** (1978) 1595.
[13] Erickson, L. E., Minkevich, I. G., Eroshin, V. K.: Biotechnol. Bioeng. **21** (1979) 575.
[14] Herbert, D.: Continuous Culture 6: Applications and New Fields, Ellis Hovwood, Chichester, 1976, p. 1.
[15] Machek, F., Štros, F., Pvokop, A., Adamek, L.: In: Continuous Culture 6: Applications and New Fields, Ellis Horwood, Chichester, 1976, p. 135.
[16] Martin, G. A., Hempfling, W. P.: Arch. Microbiol. **107** (1976) 41.
[17] Minkevich, I. G., Eroshin, V. K.: Fol. Microbiol. **18** (1973) 376.
[18] Monod, J.: Recherches sur la croissance des cultures bacteriennes, 2nd edn., Hermann, Paris, 1942.
[19] Pirt, S. J.: Proc. Roy. Soc., London, Ser. B. **163** (1965) 224.
[20] Ridgeway, J. A., Jr., Lappin, T. A., Benjamin, B. M., Corns, J. B., Akin, C. (AMOCO Foods Co.): US Patent 3865691, 1975.
[21] Vernerova, J., Rosa, M. — Czechoslovakia Patent 109568, 1964.

Acta Biotechnol. 4 (1984) 3, 267—274

Sorption of Penicillium-Cellulase on Cellulose and Lignin

B. SHAWKY[1], U. KASULKE[2], B. PHILIPP[2], G. SCHULZ[3] and W. HIRTE[3]

[1] National Research Centre, Laboratory of Microbial Chemistry, Kairo, Egypt
[2] Institut für Polymerenchemie „Erich Correns“ der Akademie der Wissenschaften der DDR
[3] Humboldt-Universität Berlin, Sektion Nahrungsgüterwirtschaft und Lebensmitteltechnologie, Wissenschaftsbereich Mikrobiologie

Summary

Sorption of Penicillium cellulase onto cotton linters samples differing in physical structure and onto KLASON-lignin from spruce has been investigated by determining cellobiase, CMCL- and "linters-activity" as well as protein content of a cellulase culture filtrate before and after contact with the sorbent at different temperature for different time intervals. CMCL- and cellobiase-activity were found to be reduced much less by the sorption process onto the cellulose samples than "linters-activity" and protein content. Sorption increased in the order: untreated linters < mercerized linters < wet fibrillated ("colloidal milled") linters. With the KLASON-lignin also a considerable sorption of the cellulase system was observed. Results are discussed with respect to preferential sorption of enzyme components.

Introduction

Sorption of the cellulolytic enzyme system onto the accessible surface of the substrate is the first step of reaction in the heterogeneous enzymatic hydrolysis of cellulose and lignocellulose. Although the important role of this enzyme sorption in the course of the whole process is wellknown and generally accepted today, information hitherto published on the equilibrium and the kinetics of this sorption step is still rather scarce as compared to the data available on rate and yield of enzyme catalyzed chain cleavage itself. Relevant works on sorption of cellulolytic enzymes to the substrate, published mainly by MANDELS and coworkers, and by GHOSE and coworkers is reviewed in detail in [1]. The amount of enzyme sorbed was found to depend largely on enzyme to substrate ratio and on temperature [2, 3] as well as on the particle size of the substrate [2], while a change of pH within a limited range seems to play a minor part only [3]. According to [3], sorption complies to a LANGMUIR isotherm, and — as reported in [2] — proceeds rather quickly within the first 20 minutes of contact between enzyme solution and cellulose, then slowing down, but continuing slowly during 8 h of contact.
In [2, 4] a stronger sorption of the "C_1-component" as compared to the "C_x-component" has been reported, while according to [5] the relative amount sorbed of these two components depends significantly on pH. In a model of heterogeneous enzymatic hydrolysis of cellulose a sorption only of "C_1 and C_x", but not of 1,4-β-glucosidase ("cellobiase") is assumed. In most of the relevant works on sorption of cellulases a rather pure cellu-

lose has been applied as a substrate. Sorption data on bagasse are reported in [6], while no information on cellulase sorption to isolated lignin was found in the published work available to us.
In this contribution, our investigations on sorption of cellulase from *Penicillium* to cellulose samples of different physical structure and to an isolated spruce lignin (KLASON-lignin) will be summarized, stressing the different effect of sorption on CMCL-activity, "linters activity", cellobiase activity and protein content of the culture filtrate applied, and thus contributing to a better understanding of the specific sorption of the various components and component groups of the enzyme system, respectively.

Experimental

As a cellulose substrate, acetate grade cotton linters ($DP_{Cu} \sim 1600$) were used, this material being applied in the air-dry, water swollen and in the mercerized (treatment with 18% aqueous NaOH solution, neutralization with aqueous acetic acid, washing and air-drying at room temperature) state, as well as after intense fibrillation in aqueous suspension ("colloidal milling"). KLASON-lignin from spruce was prepared by treatment of spruce sawdust with sulphuric acid according to [7].
Our sorption experiments were performed with culture filtrates of *Penicillium citrioviride* HU Pc 2712 [8], characterized by the following activity data:

- cellobiase activity (glucose determined by glucoseoxidase/peroxidase method, [9])
- CMCL-activity ("CMC liquifying activity": decrease of viscosity of a 1% Na-CMC-solution by action of the 1 : 100 diluted enzyme solution [9])
- "linters activity"
- protein content according to [10]

"Linters activity" as a measure of overall efficiency of the enzyme system in degrading highly crystalline cellulose was obtained by determining the weight loss of air-dry acetate grade linters and the RS (reducing sugar)-content of the hydrolysate after an incubation time of 48 h at 40 °C and pH = 5 at a substrate to liquid ratio of 1 : 100. Furtheron, the degree of polymerization (DP) of the residue was determined.
In our sorption experiments, 2.5 g (dry matter) of the appropriate sorbent were dispersed in 50 ml of the culture filtrate at a temperature of 20°, 40°, or 60°C, resp., and left without stirring at the appropriate temperature for 10, 60, or 360 min, resp. Then the liquid was separated by suction and used for the activity determinations mentioned above. As an additional information, the DP of the linters-based sorbents was assessed after the sorption process and subsequent washing and drying. Results of the activity determination after sorption were compared with the appropriate blank values of the culture filtrate before sorption, applying the necessary corrections for glucose or RS formed during the sorption process itself. In the experiments with water-swollen and with "colloidal milled" linters the dilution effect of water hold by the sorbent was taken into account by determining an appropriate blank value.

Results

Results of our experiments with air-dry linters are summarized in Table 1, which contains the appropriate activity data of the enzyme solution before and after sorption as well as the percentages of activity decrease due to enzyme sorption. The figures given for glucose (cellobiase activity) and RS (linters activity) are corrected with regard to glu-

Table 1. Sorption of cellulase (*Penicillium citrioviride*) onto untreated linters

Condition of sorption		Cellobiase activity		CMCL-activity	Protein content		Linters degrading activity				DP of residue after 48 h	DP of sorbent after sorption
Temp. (°C)	Time (min)	µg glucose/ 0.1 ml filtrate	% of original activity	% of original activity	µg protein/ ml	% of original content	RS % sugars in solution	% of original value	Weight % of sample dissolved	loss % of original value		
20	10	306	79	92	180	37	22	76	27	66	1200	1181
	60	291	75	89	180	37	20	69	23	56	1198	1133
	360	288	75	89	180	37	13	45	22	54	1158	1100
40	10	350	91	96	250	51	28	97	31	76	1197	1034
	60	342	89	94	250	51	24	83	29	71	1182	973
	360	332	86	92	250	51	21	72	28	68	1179	968
60	10	352	91	90	250	51	19	66	31	76	1016	900
	60	350	91	70	0	0	8	28	11	27	1002	807
	360	299	78	38	0	0	6	21	7	17	995	617
Activity of original culture filtrate		386	100	100	490	100	29	100	41	100	1223	

cose resp. RS formed by degradation of sorbent, these correction amounting up to 18% of total glucose resp. RS. Time of sorption obviously plays a minor part at the conditions applied here, while an increase of temperature from 20 °C to 40 °C results in a significant lowering of the sorption effect. From the dependence of activities and protein content on sorption time at 60 °C a rather severe degradation of the enzyme system even within the first hour of sorption may be concluded, in agreement with a control experiment (enzyme solution without sorption kept at 60 °C for different time intervals). This results are also in agreement with our earlier investigations on inactivation and denaturation of a cellulase solution at 60 °C, where a rather severe loss of activity of exoglucanase and non-statistically degrading endoglucanase had been stated while loss of activity of cellobiase was rather small. [10]. Thus, the further series were run at 20 °C and 40 °C only, and the data obtained at 60 °C will not be discussed in more detail.

Comparing the percentages of decrease of the different activities due to sorption at fixed external conditions, it can be stated that "linters activity" is lowered much more than cellobiase and CMCL-activity. The latter one is affected to a very small extend only. Considering once more the decrease in linters activity, RS content is obviously reduced much less than weight loss by enzyme sorption. Remarkable high — as compared to decrease in enzyme activities — is the effect of sorption on protein content of the culture filtrate. At a sorption temperature of 60 °C, no protein at all is indicated with the method employed here, but still a high cellobiase activity, a considerable CMCL-activity and some residual linters activity is found in the liquid phase after sorption.

Rather interesting results were obtained from the DP data of the sorbent after the sorption process as well as of the residues in determining "linters activity". The DP of the sorbent decreased significantly with increasing time and temperature of sorption. At 60 °C, the DP dropped from $\sim$800 to $\sim$600 between 60 min and 360 min of sorption time in spite of a protein content of nil being indicated during this time interval. These findings agree with the fact already mentioned of a rather high residual CMCL-activity even after prolonged contact between sorbent and enzyme solution at 60 °C. Somewhat unexpected at the first sight are the residue DP's obtained in linters activity determination: After sorption, i. e. after reduction of all the activities assessed here, a lower DP is found as compared to the control experiment without sorption. These residue DP's obtained with the enzyme solution after the sorption experiment show very little variation with time and temperature of sorption, and are all situated in the rather small range between 995 and 1200. An interpretation of these facts seems to be feasible on the basis of the rather large differences in sorptive binding of the various components of the cellulase system, as will be discussed later.

Sorption of cellulase onto mercerized linters (Table 2) shows the same general feature as already described for the original linters sample with regard to dependence of enzyme sorption on sorption time and temperature and with regard to the differences in decrease of the various activities due to sorption. Percentage of decrease of the various enzyme activities generally exceeds that observed with untreated linters as a sorbent (Table 1), while the order of activities with respect to this percentage remains the same. The DP values of the residues in determining linters activity after sorption are situated in the same range as with untreated linters.

From Table 3, comparing sorption of water-swollen linters before and after colloidal milling, the significant enhancement of enzyme binding due to wet fibrillation is clearly visible, while the order of the different activities with respect to the effect of sorption here also remains the same as with air-dry, untreated linters. Except the cellobiase activity, where no enhancement of enzyme binding by colloidal milling is observed. Considering this data, however, the very large correction term of 40% must be taken into account which lowers the reliability of the calculated cellobiase activity after sorption.

Table 2. Sorption of cellulase (*Penicillium citrioviride*) onto mercerized linters

Condition of sorption		Cellobiase activity		CMCL-activity	Protein content		Linters degrading activity				DP of residue after 48 h	DP of sorbent after sorption
Temp. (°C)	Time (min)	μg glucose/ 0.1 ml culture filtrate	% of original activity	% of original activity	μg protein/ ml	% of original content	RS % sugars in solution	% of original value	Weight % of sample dissolved	loss % of original value		
20	10	299	78	86	180	35	15	56	25	63	1192	938
	60	251	66	73	180	35	10	37	15	38	1185	754
40	10	338	88	96	250	49	17	63	28	70	1168	853
	60	235	85	89	250	49	16	59	25	63	1166	742
Activity of original culture filtrate		383	100	100	510	100	27	100	40	100	1220	

Table 3. Sorption of cellulase (*Penicillium citrioviride*) onto water swollen linters (10 min; 20°C)

Cellulose sample	Cellobiase activity		CMCL-activity	Protein content		Linters degrading activity			DP of sorbent after sorption
	μg glucose/ 0.1 ml culture filtrate	% of original activity	% of original activity	μg protein/ ml	% of original content	Weight % of sampl dissolved	loss % of original value	DP of residue after 48 h	
Water swollen linters	173	81	91	650	78	13	34	1185	1274
Colloidal milled linters	200	94	72	225	27	4	11	1147	373
Activity of original culture filtrate after appropiate dilution	213	100	100	835	100	38	100	1282	

Our experiments with KLASON-lignin from spruce are summarized in Table 4. Also with this non-cellulosic material a considerable sorptive binding of enzyme components is indicated by the decrease of the various activities and the protein content.
Cellobiase and CMCL-activity are reduced to about the same extend as with air-dry, untreated linters. Linters activity is decreased much less with lignin as compared to a cellulosic sorbent, and the residue DP seems to be somewhat enhanced. It can further be concluded from the data in Table 4 that the decrease of the different cellulolytic activities is not significantly lowered by changing the sorption temperature from 20 to 40 °C in some cases even an increase of sorptive binding with increasing temperature has been observed.

Discussion and Conclusions

The results obtained in this investigation confirm the strong sorptive binding of proteins of the cellulase system onto cellulose and the loss of cellulolytic activity of the liquid phase due to this sorption process. The overall dependence of sorption on time of content and temperature is also in agreement with results already published [2]. The same holds true for protein sorption: We arrived at figures of 6 ··· 12 mg protein bound to 1 g of cellulose, while PEITERSEN [3] found with his system at the conditions applied there 19 ··· 49 mg per g cellulose. Differences in enzyme sorption between the cellulose samples employed, especially between water-swollen linters before and after intense wet desintegration (colloidal milling) can be understood as being due to differences in surface area accessible to the active enzyme components [3]. Thus, the strong dependence of rate and yield in enzymatic hydrolysis of cellulose on "fibrillar morphology", mentioned in our recent publications [12, 13], can be traced back to differences in the amount of active enzyme components bound by sorption to the substrate surface.
In all of our sorption experiments, the percentage of decrease in protein content exceeds that of activity loss due to sorption, and even at a "zero protein content" indicated by the analytical method employed after prolonged sorption at 60 °C, a considerable residual cellulolytic activity was found in the liquid phase. Obviously, a rather large part of the total protein is not relevant to enzymatic chain cleavage and can be removed and/or denaturated without impairing significantly the cellulolytic activity.
These "non relevant" proteins can be considered to originate from the culture medium of *Penicillium* and from components of the enzyme system showing side activities (hemicellulases, proteases). All the enzyme activities considered here, i. e. cellobiase-, CMCL- and linters activity, are abviously reduced by the sorption process, in contrast to the model proposed by DWIVEDI and GHOSE [14] assuming a homogeneous interaction between 1,4-β-glucosidase and soluble oligomeres without sorptive binding of cellobiase to the solid surface.
But, according to our results, the relative activity decrease due to sorption is different for the various activities, linters activity being affected more than CMCL or cellobiase activity.
This is in agreement with results of MANDELS [2], who stated a stronger sorption of the C_1-component as compared to the C_x-component of cellulase. By these differences in sorptive binding of enzyme components the balance between endoglucanases, exoglucanases and 1,4-β-glucosidases is disturbed in favour of endoglucanase and cellobiase components. This may be the cause for the lowering of residue DP in linters activity determination subsequently to sorption in spite of the enhanced percentage of residue. It can be assumed that a rather large number of cellulose chains is cleaved statistically into smaller, but still insoluble fragments, resulting in a decrease of residue DP. This

Table 4. Sorption of cellulase (*Penicillium citrioviride*) onto Klason-lignin

Condition of sorption		Cellobiase activity		CMCL-activity	Protein content		Linters degrading activity				DP of residue after 48 h
							RS		Weight	loss	
Temp. (°C)	Time (min)	μg glucose/0.1 ml culture filtrate	% of original activity	% of original activity	μg protein/ml	% of original content	% sugars in solution	% of original value	% of sample dissolved	% of original value	
20	10	321	84	100	450	88	26	96	39	98	1238
	60	318	83	99	370	73	20	74	36	90	1237
40	10	321	84	97	370	73	23	85	37	93	1251
	60	309	81	89	300	59	20	74	35	88	1273
Activity of original culture filtrate		383	100	100	510	100	27	100	40	100	1230

is in agreement with the aforementioned assumption of a preferential sorption of exoglucanases and non-statistically cleaving endoglucanases.
Our sorption experiments with spruce KLASON-lignin, of course, can be considered as model experiments only, as compared to the lignin originally present in the lignin-carbohydrate moiety spruce wood, as this KLASON-lignin is severely changed in its chemical and physical structure. Nevertheless, it may be assumed that lignin plays some active part in enzyme sorption onto wood and lignocellulose and all the three types of cellulase activity considered here are obviously affected by this interaction between lignin and the enzyme system. Furthermore, the possibility can be considered that the different effect of different methods of delignification [15] on rate and yield in a subsequent enzymatic hydrolysis is caused not only by differences in cellulose accessibility, but also by differences in sorptive binding, of enzyme components to the residual, chemically modified lignin. Further experimental work will, of course, be necessary to prove the validity of this assumption.

Eingegangen: 16. 2. 1983

Literature

[1] LEE, H.-Y., FAN, L. T., FAN, L.-S.: Adv. Biochem. Eng. **17** (1980), 131.
[2] MANDELS, M., KOSTICK, J., PARIZEK, R.: J. Polymer Sci., Part. C **36** (1971), 445.
[3] PEITERSEN, N., MEDEIROS, J., MANDELS, M.: Biotechnol. Bioeng. **19** (1977), 1091.
[4] WILKE, C. R., METRA, G.: Biotechnol. Bioeng. Symp. **5** (1975), 253.
[5] GHOSE, T. K., BISARIA, V. S., DWIVEDI, C. P.: In: Proceed. V. Int. Ferment. Symp., H. Dellweg (Ed.). Berlin 1976, 439.
[6] GHOSE, T. K., BISARIA, V. S.: Biotechnol. Bioeng. **21** (1979), 131.
[7] TAPPI Standard Method T 13 m.
[8] HIRTE, W. F., GLATKE, I.: WP C 12 d/189403.
[9] SCHULZ, G.: Laboratory manual, HU Kleinmachnow 1980. Publication in preparation.
[10] SCHULZ, G.: Unpublished results.
[11] ITZHAKI, R. F., GILL, D. M.: Analyt. Biochem. **9** (1964), 401.
[12] DAO CONG DAN, JACOPIAN, V., KASULKE, U., PHILIPP, B.: Acta Polymerica **31** (1980), 388.
[13] DAO CONG DAN, FINK, H.-P., KASULKE, U., PHILIPP, B.: Acta Polymerica **32** (1981), 164.
[14] DWIVEDI, C. P., GHOSE, T. K.: J. Ferment. Technol. **57** (1979), 15.
[15] JACOPIAN, V., PHILIPP, B.: Zellstoff und Papier, to be published.

Please order the reprints from B. PHILIPP

Acta Biotechnol. 4 (1984) 3, 275—278

Short Communications

Dynamic Processing in Biotechnical Systems

B. HEINRITZ, T. BLEY, G. ROGGE, E. STICHEL and M. RINGPFEIL

Academy of Sciences of the G. D. R.
Institute of Technical Chemistry
DDR-7050 Leipzig, Permoserstraße 15

Summary

The adaptive feeding of nutrients especially of carbon and energy sources according to the demand of cells in different cell states during continuous or semicontinuous cultivation is called dynamic processing.
The deduction of dynamic process control concepts is aimed at improving the efficiency of biological substrate conversions into cell mass and other reaction products.
A growing number of results allows us to postulate that the principle of dynamic processing should be generally applied in biotechnical processes independent of the type of cells and substrates as well as the nature of products.
The basis of the deduction of dynamic process control concepts is the exact knowledge of the dependences of efficiency and rate of product synthesis on cell states during the development of cell populations.

One of the aims of biotechnological research is to improve the efficiency of biological material and energy conversion into cell mass and other reaction products. To realize this aim the introduction of dynamic processing into biotechnical processes is necessary.
The starting point of the deduction of dynamic process control concepts was the well-known existence of cell states characterized by different energetic and material efficiencies [1—8]. This means that cells possess — depending on their state — different substrate and energy demands which should be determined by extrapolation from a series of values resulting from cell populations containing different percentages of cells in the desired state. The combination of phased cultivation, fed-batch and continuous cultivation methods is suitable for determining these substrate and energy demand values.
It was found that, according to the laws of chemical thermodynamics and kinetics, excess carbon substrate is respirated by cells in states of low substrate and energy demands [5, 8, 9].
Thus, the reduction of carbon substrate feeding in repetitive states of low carbon and energy demands causes a decrease in specific material consumption and heat production by synchronized cell populations.
According to computer simulations using structured two-state models [7, 10] the optimum period of changing the carbon substrate feeding rate in dynamic cell mass production correlates to a function of the doubling time of cells and the residence time of the system.

Thus, periodically changing hydrocarbon and/or carbohydrate feeding rates in continuous cultivations of synchronized yeast populations were the first examples of dynamic processing [8, 11, 12, 13].
The synchronization of yeast populations was carried out by pH-value shocks, interruption of aeration, temperature shocks, periodic changes of residence time and/or nutrient feeding rate [14].

Table 1. Dynamic processing in continuous cell mass production by yeasts or bacteria (compare [8, 11, 12])

Fermentation system	Synchronization method	D [h]	T [h]	λ [h]	ϱ [%]	$\dot{x}$ [g/kg h]	$Y_{S/X}$ [g/g]	$\Delta Y_{S/X}$ [%]
Petroleum/*Lodderomyces elongisporus*	pH-value shocks, periodic changing substrate feeding rate	0.2	3.5	12	50	2.8	1.05	13
Petroleum/*Lodderomyces elongisporus*	Periodic changing substrate feeding rate and residence time	0.25	2.8	4	100	2.8	1.0	17
Glucose/*Lodderomyces elongisporus*	Periodic changing residence time and substrate feeding rate	0.25	2.8	3	50	4.28	1.84	16
Sucrose/*Lodderomyces elongisporus*	Periodic changing carbon substrate feeding rate	0.25	2.8	3	50	4.1	1.96	16
Sucrose/molasses/*Lodderomyces elongisporus*	Periodic changing carbon substrate feeding rate	0.28	2.5	3	50	4.2	1.44	20
Sucrose/paraffin/*Lodderomyces elongisporus*	Periodic changing carbon and energy feeding rates adding paraffins during phase of reduced sucrose feeding according to the optimum carbon/energy ratio of substrates for cell mass production	0.25	2.8	3	50	4.3	1.9	20
Glucose/*Saccharomyces cerevisiae*	Periodic changing carbon substrate feeding rate	0.25	2.8	3	50	4.2	1.9	14
Glucose/*Escherichia coli*	Periodic changing carbon substrate feeding rate	0.67	1.0	1.2	50	4.1	1.8	15

The knowledge gained in the field of cell mass production by yeasts was applied to bacterial cultivations on carbohydrates.
All these results show an improvement of the specific material consumption and heat production coefficients up to 20% by dynamic processing compared to steady-state fermentation (Table 1).
But dynamic processing can also be suitable for the economic production of intra-cellular as well as extra-cellular products such as lipids, polysaccharides, polyphosphates, enzymes, organic acids or alcohols.
It is advantageous to distinguish between cellular products which are synthesized during the proliferation and the post-proliferation state of cells (Table 2).

Table 2. Dynamic processing in synthesis of intra-cellular and extra-cellular products by yeasts and bacteria including cell doubling under near optimum conditions

Product	Fermentation system	Synchronization method	Cell state of product synthesis and harvest	Reference
Citric/iso-citric acid	Paraffin/*Candida lipolytica*	Exchange of fermentation medium in intervals, periodic exhaustion of ammonium nitrogen in fermentation medium	Product synthesis in post-proliferation state; product harvest at a time when acid production rate is maximum	[15, 16]
Ethanol	Sucrose/*Saccharomyces cerevisiae* H 155	Changes between aerobic and anaerobic conditions and exchange of fermentation medium in intervals	Product synthesis in proliferation state; product harvest when ethanol production rate is maximum	[17, 18]
Cell mass of a high crude protein content	Glucose/*Lodderomyces elongisporus* DEH 15	Exchange of fermentation medium in intervals including periodic cell doubling	Product synthesis in proliferation state, product harvest at a time when crude protein content is maximum	[19]
Cell mass of a high polyphosphate content	Glucose/*Lodderomyces elongisporus* DEH 15 (phosphate excess)	Exchange of fermentation medium in intervals including periodic cell doubling	Product synthesis in post-proliferation state, product harvest at a time when polyphosphate content is maximum	[19]

A common property of all the possible dynamic process control concepts for the synthesis of cellular products is the necessity of synchronizing of cell functions, the time dependent harvest of products and the carrying out of cell proliferation under near-optimum conditions.
The dynamic process control concepts consist preferentially of procedures like

- semicontinuous cultivation of cell populations, including cell doubling under near-optimum conditions,
- changes of nutrient concentrations, pH-value shocks and/or temperature shocks for synchronization,
- adaptive nutrient feeding according to the material and energy demands of cells,
- product harvest in fixed time intervals.

All these results summarized in tables 1 and 2 allow to postulate that the principle of dynamic processing should be of general importance in biotechnical processes independet of the type of cells and substrates as well as the nature of products.
This principle consists in the adaptive feeding of material and energy in continuous or semicontinuous cultivations of synchronized cell populations.
An important prerequisite for realizing theoretical material and energetic efficiencies of biological substrate conversion into cell mass and other reaction products is the choice of optimum conditions for cell proliferation.

Symbols

D	Flow rate
T	Doubling time of cells
λ	Optimum period of changing carbon substrate feeding rate
ϱ	Amplitude of changing carbon substrate feeding rate (substrate feeding rate of steady-state 100%)
$\dot{x}$	Cell mass production rate
$Y_{S/X}$	Specific substrate consumption coefficient
$\Delta Y_{S/X}$	Increase of specific substrate consumption coefficient compared to steady-state

Eingegangen: 31. 3. 1983

References

[1] MEYENBURG, K. v.: Katabolitrepression und der Sprossungszyklus von *Saccharomyces cerevisiae*. Dissertation B, Zürich 1969.
[2] DAWSON, P. S. S.: J. Appl. Chem. Biotechnol. **22** (1972), 79.
[3] STREIBLOVA, E.: Biol. Rundsch. **18** (1980), 350.
[4] RABOTNOVA, I. L.: Kultivirovanije mikroorganismov. Akademija nauk, Moskva SSSR 1981.
[5] HEINRITZ, B.: Dissertation A, AdW der DDR, Leipzig 1978.
[6] BLEY, T., HEINRITZ, B., STEUDEL, A., STICHEL, E., GLOMBITZA, F., BABEL, W.: Z. Allg. Mikrobiol. **20** (1980), 275.
[7] BLEY, T.: Dissertation A, Leipzig 1981.
[8] HEINRITZ, B., ROGGE, G., STICHEL, E., BLEY, T.: Acta biotechnol. **3** (1983), 125.
[9] GLOMBITZA, F., HEINRITZ, B.: Z. Allg. Mikrobiol. **19** (1979), 171.
[10] BLEY, T., SCHMIDT, A.: Studia biophysica **18** (1980), 11.
[11] BLEY, T., GLOMBITZA, F., HEINRITZ, B., STICHEL, E.: Wirtschaftspatent DD-WP 140 150, Anmeldetag 5. 10. 80.
[12] HEINRITZ, B., STICHEL, E., BLEY, T., ROGGE, G., GLOMBITZA, F.: Z. Allg. Mikrobiol. **21** (1981), 581.
[13] RINGPFEIL, M.: Proc. 2. Leipziger Biotechnologiesymposium 1982.
[14] STICHEL, E., ROGGE, G., BLEY, T., HEINRITZ, B.: Z. Allg. Mikrobiol. **22** (1982), 717.
[15] ISKE, U., BULLMANN, M., GWENNER, CH., HEINRITZ, B.: Proc. Jahrestagung der Biologischen Gesellschaft der DDR, Reinhardsbrunn 1981.
[16] BULLMANN, M.: Diss. A. In preparation.
[17] HEINRITZ, B., GHILDYAL, N. P., RINGPFEIL, M.: Z. Allg. Mikrobiol. In press.
[18] HEINRITZ, B., ROGGE, G., HANSCHMANN, G., RINGPFEIL, M.: Acta biotechnol. In press.
[19] HEINRITZ, B., ROGGE, G., STICHEL, E., BLEY, T.: Unpublished results.

Acta Biotechnol. 4 (1984) 3, 279—280

Stirrer and Peristaltic Pump Agitated Fermenter

H. E. NIKOLAJEWSKI, L. MEINEKE and G. WETTIG

Staatliches Institut für Immunpräparate und Nährmedien
1120 Berlin, Klement-Gottwald-Allee 317—321

Summary

A fermenter has been designed for aerobic cultivation processes with stirrer and external peristaltic pump to provide high oxygen transfer rate along with low foam formation. Aerobic growth experiments conducted with *E. coli* gave an oxygen transfer to ensure a maximum growth rate of about 9 g dry weight/l h. In all tests performed foam never reached a critical level.

The patented fermenter with stirrer and peristaltic pump agitation [1] was specifically designed to ensure both a high oxygen transfer efficiency along with low foam formation. The new fermenter system (see Fig. 1) is equipped with a stirrer and an external peristaltic pump circulating the culture medium through an external line under vigorous agitation of the liquid in the vessel. The air is introduced into the culture medium near the end of the external circulating line spraying the culture medium above the stirred culture liquid in the fermenter. The effluent gas is passed through water cooled condenser where vapour and aerosols are condensed and returned into the fermenter. Comparative cultivation experiments were conducted with *E. coli* in a 1 litre vessel containing 0.5 l of a

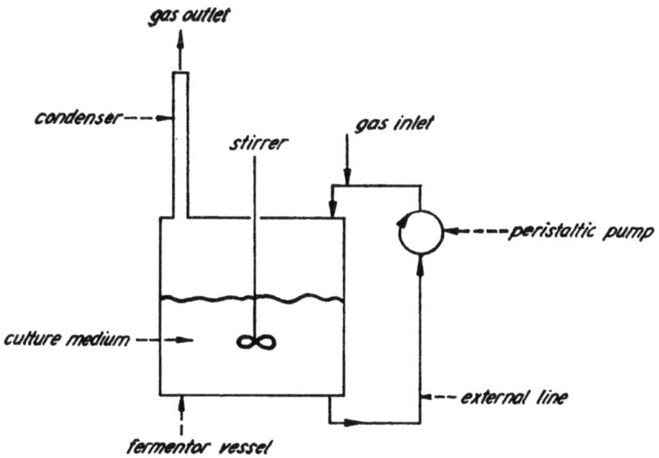

Fig. 1. Schematic diagram of the stirrer and peristaltic pump agitated fermenter

glycerol-casein hydrolysate medium adjusted to pH 7.0. In this set of experiments (Table 1) stirrer speed as well as air flow rate were kept constant while the speed of the peristaltic pump was varied over the range of 0 to 60 l/h. In addition, results obtained in above experiments were compared with growth rates and foam forming properties performing cultivation experiments in a submerged aerated fermenter.

Table 1. Comparative tests with *E. coli* using the stirrer and peristaltic pump agitated fermenter (A) and the submerged aerated fermenter (B)

fermenter type	pump speed l/h	maximum cell yield $\frac{\text{g dry weight}}{\text{l}}$	maximum growth rate $\frac{\text{g dry weight}}{\text{l} \cdot \text{h}}$	antifoam reagents addition
A	0	1.5	0.5	0.5 ml
A	10	12	3.2	none
A	20	19	4.2	none
A	30	22	6.3	none
A	60	32	9.0	none
B	none	6.6	2.2	1 ml

A: 1 l vessel with 0.5 l culture medium, stirrer speed 750 rpm and an air flow rate of 1 l/min.
B: 1 l vessel with 0.5 l culture medium, stirrer speed 750 rpm and an air flow rate of 1 l/min

The data obtained (Table 1) show clearly the unique influence of peristaltic pump agitation on the increase of growth rates and cell yields. Thus, growth rates up to 9 g/l and cell yields up to 35 g/l could be achieved while foam formation never reached a critical level whenever peristaltic pump agitation have been applied. As yet similar high growth rates have been reported by BAUER and SHILOACH [2] cultivating *E. coli* in a commonly used stirred fermenter of submerged gassed type. However, pure oxygen instead of air had to be applied to meet the demand of oxygen transfer for correspondingly high yields and growth rates.
Finally it must be emphasized that results in Table 1 are not adapted to optimal cultivation conditions. Thus, a thorough analysis of aeration efficiency depending on variation of process parameters (e.g. stirrer speed and gas flow rate) will be a matter of further investigations.

Eingegangen: 4. 4. 1983

References

[1] NIKOLAJEWSKI, H. E., MEINEKE, L., WETTIG, G.: DDR-Pat. Nr. WP C 12 B/240961.
[2] BAUER, S., SHILOACH, J.: Biotechnol. Bioeng. 16 (1974), 933.

Acta Biotechnol. 4 (1984) 3, 281–284

Biosorption of Mercury by Microorganisms

F. GLOMBITZA[1], U. ISKE[1], CHR. GWENNER[1] and M. R. V. KRISHNAN[2]

[1] Academy of Science of the G.D.R.
Institute of Technical Chemistry
7050 Leipzig, Permoserstraße 15

[2] Department of Chemical Engineering
A. C. College of Technology
Anna University, Madras-600025

Summary

Biomasses are able to sorb great amounts of mercury ions from water solutions.
Sorbed concentrations are dependent on the pH-value, the strain of microorganisms and the substrate used for cultivation.
The uptake of mercury takes place at a high reaction rate and is completed after few minutes.
A comparison between maximum amounts of uptake concentration and theoretical maximum amounts with a total saturation of the surface leads to the assumption that mercury is bound to definite sites.

Aim

SHIE and BABER [1] reported a mechanism of the uptake of mercury by microorganisms consisting of two steps. These steps are

- an uptake of mercury on the surface of microorganisms by means of a sorption on the cell wall and membranes
- an accumulation in the interior of microorganisms.

BRUNKER [2] and BROWN [3] located mercury in yeasts and found mercury in mitochondria, membranes of the nucleus and in endoplasmic reticula.
The uptake of mercury by means of microorganisms is an important process for removing mercury and its compounds from waste water and waste materials.
If the uptake takes place during the growth of microorganisms the microorganisms are able to form volatile organic mercury compounds [4].
Because a recycling of mercury would require the extraction of all the air in the latter case the first step of the uptake reaction is investigated, analyzed and used for removing mercury from waste water.

Methods

Different biomasses were mixed with a mercury ion solution and separated again after a chosen mixing time.
The residual concentration of mercury in the solution was determined after separating it from the biomass.

This mixing reaction was repeated with a fresh solution and the same biomass until a non-decreasing constant mercury ion concentration in the solution was observed.
The solution was a mercury II chloride solution with a concentration of 300 mg/l (300 mg Hg^{2+}).
The maximum mercury concentration taken up was calculated from the number of the sorption reactions and the difference of mercury concentration between the successive reactions.
The biomass used were yeasts and bacteria cultivated with different substrates, partly resting microorganisms or denatured dead microorganisms.

Results

Figure 1 shows the sorption lines for different biomasses and different numbers of reactions till a saturation of the microorganisms with mercury was observed.
Bacteria cultivated on methanol show the highest specific amount of mercury taken up.

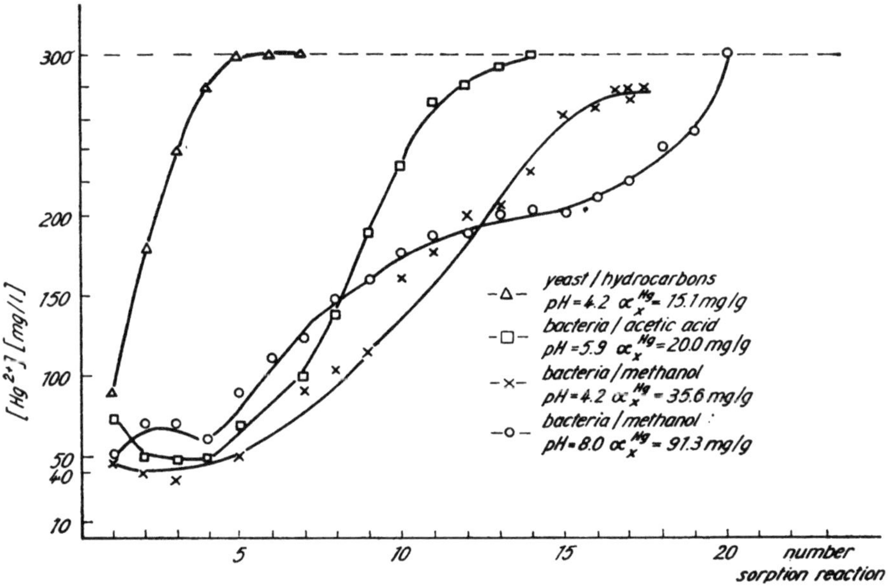

Fig. 1. Dependence of mercury concentration on the number of sorption reactions before saturation of biomass with mercury and maximum mercury concentration taken up

Table 1 contains different biomasses and maximum values of mercury concentrations taken up for a range of pH from 4.9 to 5.8.
It may be concluded from these results that

- no distinction exists between living resting and dead cells with respect to the uptake of mercury ions
- the capability to take up mercury decreases due to a treatment of the biomass with acetone
- bacteria cultivated with methanol exhibit the highest specific uptake concentrations.

Table 2 shows the mercury ion concentration taken up as a function of pH.
These results demonstrate that

- all biomasses show a high capability of sorption in a pH range near 8

Table 1. Maximum concentration of mercury ions taken up by differently treated microorganisms

microorganisms/ substrates	pH	mercury/ uptake [mg/g]	pretreatment of biomasses
bacteria			
acetic acid	5.3	25	freeze-dried
dto.	5.4	25.7	resting cells
bacteria			
methanol	4.9	46.0	freeze-dried
yeasts			
carbohydrates	5.8	41.4	freeze-dried
bacteria			
hydrocarbons	5.7	36.0	dried in spray drier
yeasts			
hydrocarbons	5.0	38.0	freeze dried
yeasts			
hydrocarbons	5.3	22.5	washed with acetone
yeasts			
hydrocarbons	5.8	26.5	washed with acetone and pulverized

- in an acidic pH range (1.7—2.0) no sorption was observed
- bacteria cultivated on acetic acid or methanol possess a higher specific capability of sorption than yeasts cultivated on hydrocarbons and washed with acetone to remove residues of hydrocarbons.

Of great importance is the information about sorption reaction rates. Therefore the biomasses were mixed with mercury ion containing solutions and separated again after a certain time.
The residence time of the biomasses in solution was continuously shortened.
The results are shown in Figure 2.
The figure shows a high reaction rate of sorption.
The mercury uptake is completed after few minutes.

Table 2. Dependence of maximum mercury ion concentration taken up on pH-value for different biomass

microorganisms — substrate	pH / α_x^{Hg} specific uptake concentration						
bacteria	pH	1.7	2.0	5.3	5.9	7.7	7.9
acetic acid	α_x^{Hg}	—	—	25	20	109.3	78.7
bacteria	pH		4.2		4.9		8.0
methanol	α_x^{Hg}		35.6		46.0		91.3
yeasts	pH		4.2		5.3	6.0	8.0
hydrocarbons	α_x^{Hg}		15.1		22.5	6.7	40.7

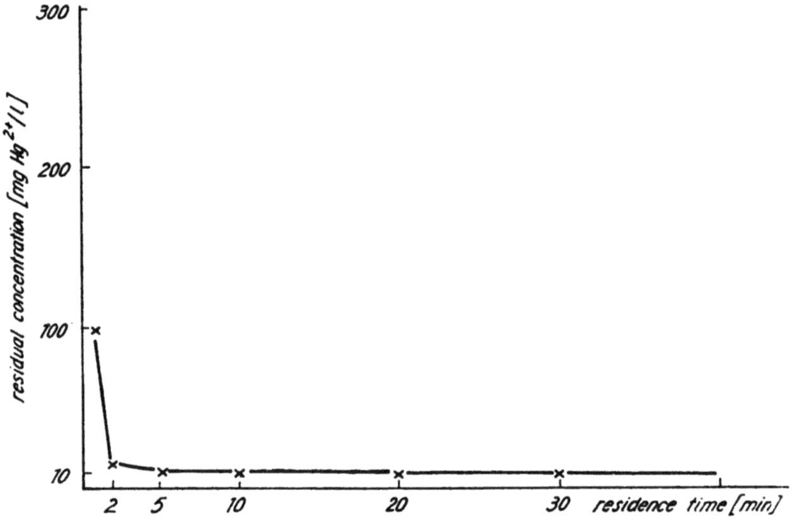

Fig. 2. Dependence of residual mercury II ion concentration on residence time of microorganisms in suspension

Discussion and Conclusions

If we assume a sorption of mercury on the surface of microorganisms we have to answer the question concerning the theoretical maximum value of mercury.
For this we assume a yeast cell with a size of 3×5 μm and the shape of a cylinder.
The surface area of this cell is 61.23 μm^2 and 1 g of biomass consists of 2.8×10^{14} cells.
Mercury chloride is not dissociated (only a small part) and the diameter of a molecule is 11.5 Å.
We assume further a distance between neighbouring molecules of 2 Å.
Then 5.96×10^7 molecules may be arranged on the surface of such a cell as a monatomic layer.
This means a mass of 5.6 g mercury is sorbed by 1 g of yeast.
This value is very high and lies above the experimentally found results.
If we assume a structured surface then we have a greater surface area for sorption.
Therefore we have to assume a sorption of mercury on definite sites of surface.
This is confirmed by results of Roshart [5], who observed a sorption on S and O groups.

Eingegangen: 5. 5. 1983

References

[1] Shie, Y. J., Baber, J.: Planta 109 (1973), 49.
[2] Brunker, R. L.: Appl Microbiol. **27** (1974), 870.
[3] Brown, T. A.: J. Gen. Microbiol. **99** (1977), 435.
[4] Lander, L.: Nature **230** (1971), 452.
[5] Roshart, R. G.: Water Pollut. Res. Canada **9** (1974), 62.

Acta Biotechnol. 4 (1984) 3, 285—288

Cultivation of Mercury Resistent Microorganisms

F. Glombitza, U. Iske and Chr. Gwenner

Academy of Science of the GDR
Institute of Technical Chemistry Leipzig
7050 Leipzig, Permoserstraße 15

Summary

A strain of bacteria was isolated from soil samples. The resistance of these microorganisms to mercury ions was proved by means of discontinuous and continuous fermentations.
Continuous fermentation was carried out also at a mercury ion concentration of 250 mg/l.
A balance of mercury distributions demonstrates the capability of these bacteria to accumulate a great amount of mercury.

Aim

It is known that microorganisms are capable of removing mercury or compounds of mercury from water solutions [1].
Mercury is accumulated in microorganisms or microorganisms produce volatile organic mercury compounds [2—4].
Sing and Sherman [5] reported that the formation of methyl mercury compounds is connected with the capability of microorganisms to transfer methyl groups and their demand of methionin.
For mercury recovery such a reaction is undesirable. In addition, volatile organic mercury compounds are strong poisons.
Therefore a mercury resistent strain of microorganisms was to be isolated and characterized which accumulates mercury with high concentrations and without the formation of volatile mercury compounds.

Isolation of the Strain

Samples of soil are treated in shaker bottles using glucose as substrate and a complex mineral medium, a temperature of 30 °C and different $HgCl_2$ concentrations in the medium. The pH value is controlled in the range of 6—7.
The microorganisms grown are cultivated discontinuously under the same conditions.
The next step is a continuous fermentation. The microorganisms thus obtained are a population of bacteriae like *Acetobacter* and *Azotobacter*.

Characterization of Microorganisms

The characterization of the microorganisms for the description of the influence of mercury concentration was carried out in a reactor by means of a cultivation method after DAWSON [6]. This is the "phased culture". In it the concentration of nutritives and substrates is chosen so as to permit a doubling of the amount microorganisms. When the biomass has reached double concentration then a half of the reactor contents is exchanged with a new mineral medium and substrate.
All conditions are kept constant during each cycle.
The mercury concentration was increased and allowed us to measure the influence of the concentration on microorganism growth.
Figure 1 demonstrates this "phased culture" principle for different concentrations of mercury ions.

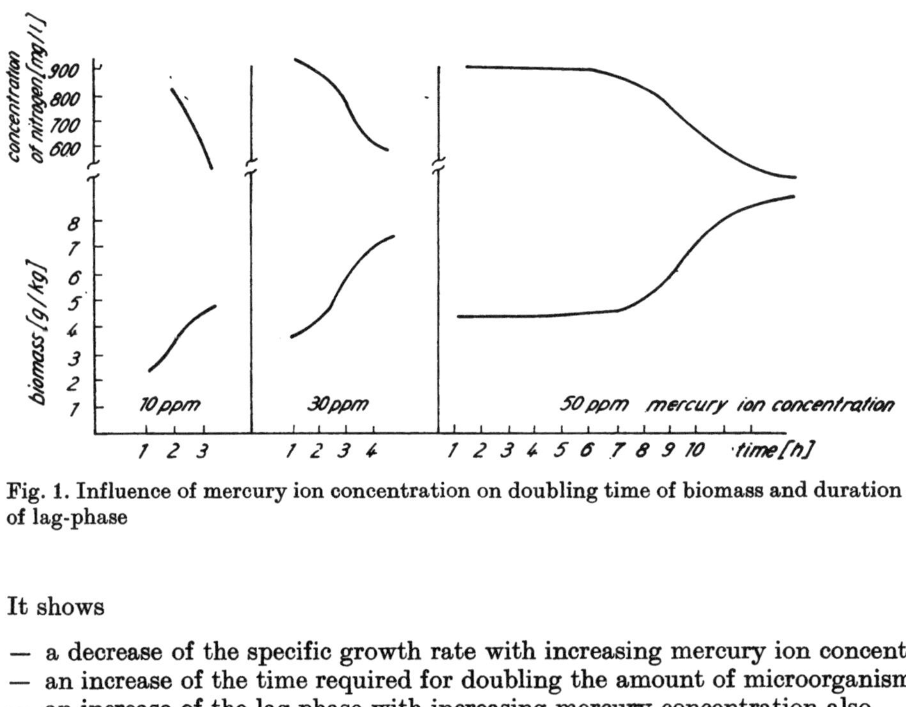

Fig. 1. Influence of mercury ion concentration on doubling time of biomass and duration of lag-phase

It shows

- a decrease of the specific growth rate with increasing mercury ion concentration,
- an increase of the time required for doubling the amount of microorganisms,
- an increase of the lag-phase with increasing mercury concentration also.

The decrease of the maximum specific growth rate is shown in Figure 2. It decreases from 0.43 h^{-1} for a mercury ion concentration of 10 mg/l to 0.2 h^{-1} for a concentration of 50 mg/l.

Continuous Fermentation

Continuous fermentation is carried out with increasing mercury ion concentration for investigating the behaviour of bacteria at higher concentrations of mercury ions.
The concentration is raised from 50 mg/l to 250 mg/l.

Conditions of Fermentation

substrate	: glucose	residence time	: 7 h
pH	: 6.6 ± 0.2	aeration	: 100 l/kg h
temperature	: 31 °C ± 1.0		

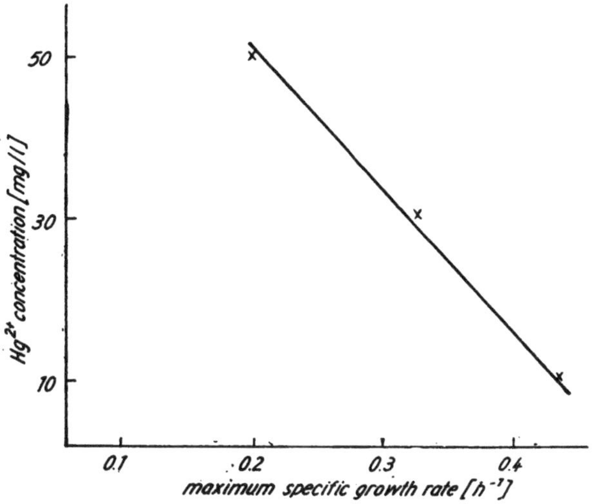

Fig. 2. Dependence of maximum specific growth rate on mercury ion concentration

Results

Fermentation results are shown in Table 1 and Figure 3.

Figure 3 shows a typical fermentation process with a mercury ion concentration of 50 mg/l.

The specific demand of glucose is constant and independent of the mercury concentration in this range.

Average data for Y_s^{-1} vary from 2.3 to 2.8 g glucose per gramme of biomass.

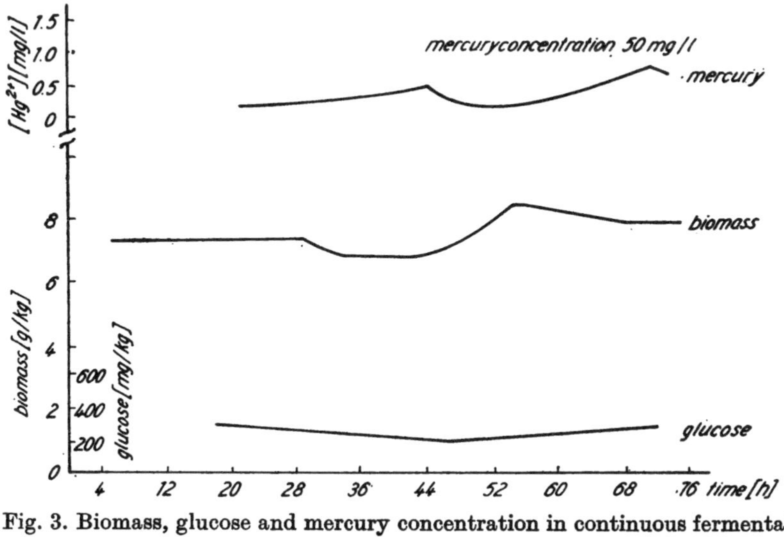

Fig. 3. Biomass, glucose and mercury concentration in continuous fermentation

Table 1. Dependence of fermentation results on mercury ion concentration

concentration of mercury	mg/l	50	100	150	200	250
biomass	g/kg	7.8	7.5	8.5	7.2	6.8
productivity	g/kg h	1.1	1.07	1.2	1.03	0.97
actual substrate concentration	mg/l	~ 400				
actual concentration of mercury ions	mg/l	≦ 1				

Balance of Mercury

For getting information about the distribution of mercury in a fermentation system the concentration of mercury was determined and a balance was formulated (Tab. 2). From this balance an accumulation of mercury in the biomass of 160 mg is calculated. This means a great amount of mercury is stored in the biomass during the growth of cells.

Table 2. Balance of mercury

input	output			
concentration of mineral medium	conc. of biomass	spec. conc. of Hg in biomass	conc. in air	conc. in broth
200 mg/l	7.2 g/kg	22.2 mg/g	15 mg/24 h	1 mg/l

Analyses of the air after fermentation (2400 l/24 h) for the determination of the concentration of volatile organic mercury compounds give 15 mg mercury only in a volume of 1 l of an extraction solution and a time of 24 h.
Existing differences between the input and output mercury concentrations may be explained by differences of the analytical methods. But the results clearly demonstrate the capability of isolated microorganisms to accumulate mercury with high concentrations.

Eingegangen: 5. 5. 1983

References

[1] Jeffries: Progr. ind. Microbiol. **16** (1982) 21.
[2] Rothstein, A.: Fed. Proc. **18** (1959) 1026.
[3] Brown, T. A.: J. Gen. Microbiol. **99** (1977) 435.
[4] Lander, L.: Nature **230** (1971) 452.
[5] Sing, A., Sherman, F.: Nature **247** (1974) 227.
Sing, A., Sherman, F.: J. Bacteriol. **118** (1974) 911.
[6] Dawson, P. S.: J. Appl. Chem. Biotechnol. **22** (1972) 79.

Acta Biotechnol. 4 (1984) 3, 289–292

Cultivation of Yeasts on Hydrolysate Nutrient Medium Obtained from the Production of Microcrystalline Cellulose

T. SOKOLOV[1], V. KOSTOV[2], R. VELKOVA[1], L. KEHLIBAROVA[2], ST. JELEV[1], B. TZVETKOVA[2], J. MLADENOVA[1] and L. LOSEVA[2]

[1] Higher Institute for Chemistry and Technology
Department "Chemical Technology of Wood"
1156 Sofia
[2] Bulgarian Academy of Science
Microbiological Institute
1113 Sofia

Summary

Studies of the biomass production during a continuous cultivation of yeasts on a nutrient medium, prepared from a hydrolysate from the production of microcrystalline cellulose, have been carried out. A new strain of yeasts has been used. Its cultivation has been achieved without addition of biostimulators to the nutrient medium in spite of their absence in the initial hydrolysate. Practically a complete assimilation of sugars has been achieved at high dilution rates ($D = 0.25$ to $0.50\ h^{-1}$). The yield of biomass achieved is above 50% compared to the initial sugars and it contains 48.89% true protein. The results obtained offer the possibility of a complex utilization of the products of cellulose hydrolysis in the production of microcrystalline cellulose with a realization of a waste free technology.

Introduction

The hydrolysates from the production of microcrystalline cellulose (MCC), obtained in the hydrolytic process in the presence of diluted mineral acids represent diluted aqueous solution of hydrolytic sugars, mainly glucose and sulfuric or hydrochloric acid. They are obtained after the hydrolysis of cellulose, isolated through chemical processing of wood. They possess low concentration of sugars and do not contain biostimulators occuring in the hydrolysates from natural wood [1]. The lack of these stimulators implies a negative influence on the development of the auxotrophic strains of yeasts utilized in the practice in course of biosynthesis of microbial protein. It is established that on nutrient medium prepared from hydrolysates received during the production of MCC, valuable development under batch cultivation is possible for such strains, which are in condition to synthesize their own growth substances, e.g. *Candida acapsuligena* n. sp. 74 — 4, *C. scottii*, Tul 6 and *C. blankii* 78 — 16 [2].
The purpose of the present work is to receive a biomass under the continuous cultivation of yeasts on nutrient medium, prepared from a hydrolysate of the production of microcrystalline cellulose.

Materials and Methods

Nutrient medium. A hydrolysate obtained during the production of MCC according to [3] with a content of 0.5% sugars and 1% sulfuric acid is used. From the latter was prepared nutrient medium by the neutralization at a temperature of 80 °C with CaO up to pH 3.5. $(NH_4)_2SO_4$, H_3PO_4 and KCl are added in quantities which ensure 6% nitrogen, 3% P_2O_5 and 0.8% K in relation to the amount of the sugars in the hydrolysate. As well for 1 l nutrient medium were added: $MgSO_4$ — 0.35 g; NaCl — 2.5g; H_3BO_3 — 500 μg; $CuSO_4 \cdot 5H_2O$ — 40 μg; KJ — 100 μg; $FeCl_3 \cdot 6\,H_2O$ — 200 μg; $MnSO_4 \cdot H_2O$ — 400 μg; $NaMoO_4 \cdot 2H_2O$ — 200 μg and $ZnSO_4 \cdot 7H_2O$ — 400 μg. The solution is cooled, neutralized with 25% ammonia water up to pH 4.5 and the precipitates are removed through filtration.

Microorganisms. Candida acapsuligena n. sp. 74 — 4 from the collection of the Institute of Microbiology of the Bulgarian Academy of Science, Sofia, PR of Bulgaria. As inoculating material fresh biomass from 48 h cultivation of malt agar 8° Balling was used.

Apparatus and conditions for cultivation. Investigation were realized in a fermenter "ANCUM-2" under continuous cultivation at a temperature of 38°C, pH 4.5 with an aeration rate of 1 vvm, agitation speed of 1000 rpm and dilution rate $D = 0.25\ h^{-1}$; $0.38\ h^{-1}$ and $0.50\ h^{-1}$.

Analytical control. The concentration of residual sugars in the culture broth were measured by the ebulliostatic method [4] and the amount of yeasts by weight as absolutely dry biomass dryed at 105 °C. The yeasts are analysed for the content of raw proteine according to KJELDAHL [4]. The amino acid composition of an average sample from dried yeasts is determined by means of amino analyser Bio Cal — BC — 200 after dialysation of the sample.

Results and Discussion

Data from the tests are presented in Table 1 and 2. A continuous cultivation of yeasts in hydrolytic nutrient medium of MCC without complementary addition of biostimulators was realized. Practically, it is achieved complete assimilation of the sugars at a high dilution rate ($D = 0.25 - 0.50\ h^{-1}$). It is prepared a yeasts biomass which has by organoleptic evaluation in dry conditions no odour, a fresh (agreeable) taste and a white colour, which is preferable in the utilization of yeasts for the preparation of protein isolates and for nutrient purposes.

From Table 1 can be seen that the yield of absolutely dry yeasts (ADY) from the initial sugars is over 50% and at the higher rates of exchange of the nutrient medium (dilution rates) the yields of biomass are higher and the yeasts have a lower protein content. The yield of protein from the sugars is 26.36—31.32%. The bulk protein content in the bone

Table 1. Results from the continuous cultivation of *C. acapsuligena* n. sp. in a nutrient medium obtained from a hydrolysate of production of microcrystalline cellulose

No	Index	Dimension	Cultivation conditions		
			1	2	3
1.	Dilution rate	h^{-1}	0.25	0.38	0.50
2.	Yield of ADY from sugars	%	50.34	61.02	57.07
3.	Content of protein in ADY	%	52.47	51.33	50.11
4.	Yield of protein from sugars	%	26.36	31.32	28.59
5.	Residual content of sugars	%	< 0.01	< 0.01	< 0.03

Table 2. Aminoacid composition of yeast biomass prepared by cultivation of *C. acapsuligena* n. sp. 74-4 in hydrolysate medium from the production of MCC

No	Aminoacid	Content of residual aminoacids		Content of aminoacid
		in the protein complex %	in ADY %	equivalent in ADY %
1.	Aspartic acid	11.81	5.77	6.67
2.	Threonine	5.62	2.75	3.26
3.	Serine	3.94	1.93	2.33
4.	Glutamic acid	11.47	5.61	6.39
5.	Proline	3.97	1.94	2.30
6.	Glycine	5.02	2.45	3.22
7.	Alanine	7.88	3.85	4.82
8.	Cysteine	2.33	1.14	1.23
9.	Valine	8.05	3.94	4.66
10.	Methionine	1.87	0.91	1.04
11.	Isoleucine	4.67	2.28	2.64
12.	Leucine	8.96	4.38	5.05
13.	Tyrosine	4.91	2.40	2.66
14.	Phenylalanine	5.98	2.93	3.29
15.	Lysine	4.65	2.27	2.59
16.	Histidine	2.25	1.10	1.24
17.	Arginine	6.62	3.24	3.57
		100.00	48.89	56.99

dry biomass is 50.11—52.47%; the true protein evaluated through the amino acid composition in ADY is 48.89% (Table 2), which stands in favour of the high content of true protein in the prepared biomass.

Data from the accomplished analysis indicates that this biomass contains 1.87% methionine, 2.25% histidine, 6.62% arginine, 2.25% valine, 5.98% phenylalanine and 5.62% threonine by the true protein. It can be related to the class of yeasts rich in irreplaceable amino acids [6].

A principal part in the achievement of the discussed results plays the choice of yeast strains used in the experiments, utilized for the first time in researches on microbiological synthesis of protein.

The utilization of hydrolysates of MCC through cultivation of yeasts leads to a bilateral advantage. From one point, a vitaminous protein product is obtained, on the other hand, the oxidizable organic compounds from waste hydrolysates are turned harmless which benefits preservation of the environment.

The data from the tests suggest an opportunity for complex utilization of the products from hydrolysation of cellulose during the preparation of microcrystalline cellulose with realization of a wasteless technology.

Eingegangen: 6. 6. 1983

References

[1] KOSTOV, V., KECHLIBAROVA, L., DENCHEV, D.: Acta microbiol. 8 (1981), 74.
[2] KOSTOV, V., SOKOLOV, T., VELKOVA, R., JELEV, S.: Compt. Rend. Acad. Bulg. Sci. 5 (1983), in press.
[3] SOKOLOV, T., GOSPODINOV, N., VIDIMSKI, E.: AC-PRB No. 19462 (1973).
[4] EMELIANOVA, I. Z.: Chemical and Technical Control of Hydrolysates Production, Wood Industry, Moskow, 1976, pp. 33, 227.
[5] ROZE, A. H., HARRISON, J. S.: The Yeasts, London, New York: Academic Press, 1970, p. 439.
[6] TOMME, M. F., MARTINENCO, P. B.: Aminoacid Composition of Fodders, Moscow: Kolos, 1972, p. 255.

Acta Biotechnol. 4 (1984) 3, 293—296

Untersuchungen zur Anreicherung von Ubichinon-9 aus Lipid-Kohlenwasserstoff-Fraktionen mittels Kurzwegdestillation

B. VOIGT[1], H. MÜLLER[1], M. WORBS[1], F. WINKLER[2] and U. KÖHLER[2]

[1] Akademie der Wissenschaften der DDR
Institut für technische Chemie Leipzig
7050 Leipzig, Permoserstraße 15
[2] VEB Chemieanlagenbaukombinat Leipzig/Grimma
Technikum Holzhausen
7124 Holzhausen, Ernst-Thälmann-Straße 63

Summary

Short path distillation was used for the enrichment of ubiquinone-9 from a lipid hydrocarbon extract. The main components of the lipid hydrocarbon extract were hydrocarbons of gas oil (b. p. 513 to 653 K), phosphatides, glycerides, and fatty acids. Phosphatides were isolated by extraction with acetone before distillation. The acetone soluble fraction was distilled by a pressure of $\leqq 0.5$ kPa and a temperature of 433 to 453 K. The concentration of the ubiquinone-9 in the residue of distillation was three times higher than in the acetone soluble fraction.
Ubiquinone-9 can be obtained by a simplified separation from the enriched fraction.

Ubichinone sind in aerob kultivierten Mikroorganismen im Vergleich zu anderen biologischen Ausgangsmaterialien in relativ hoher Konzentration enthalten [1]. Für die Ubichinonbildung ist ein Kohlenwasserstoff-Substrat besonders günstig [2]. Die aus Kohlenwasserstoff-utilisierenden Mikroorganismen gewonnenen Lipid-Kohlenwasserstoff-Fraktionen stellen somit ein geeignetes Ausgangsmaterial für die Gewinnung der Ubichinone dar. Eine Isolierung ist durch den hohen Anteil an Kohlenwasserstoffen jedoch besonders aufwendig. Wir untersuchten deshalb, inwieweit sich die Gewinnung der Ubichinone vereinfachen läßt, wenn die Kohlenwasserstoffe teilweise destillativ abgetrennt und gleichzeitig die Ubichinone im Sumpf angereichert werden.
Ubichinone (2,3-Dimethoxy-5-methylbenzochinone mit variabler terpenoider Seitenkette) sind licht-, luft- und alkaliempfindliche hochmolekulare Substanzen. Für eine Anreicherung wurde deshalb die Kurzwegdestillation gewählt, die sich bei der Destillation hitzeempfindlicher, hochmolekularer Verbindungen bewährt hat.
Das Verhalten von Ubichinonen unter den Bedingungen einer Kurzwegdestillation ist bisher nicht bekannt.
Phosphatide sind nicht destillierbar und müssen vor der Destillation aus der Lipid-Kohlenwasserstoff-Fraktion entfernt werden [3].

Material und Methoden

Für die Untersuchungen wurde ein Modellgemisch folgender Zusammensetzung eingesetzt:

31,0 Gew.-% Phosphatide
13,8 Gew.-% Glyceride und freie Fettsäuren

53,5 Gew.-% Kohlenwasserstoffe (Kp 473—653 K)
0,4 Gew.-% Ubichinon-9
1,3 Gew.-% Ergosterol

Das Modellgemisch entspricht in seiner Zusammensetzung etwa der Lipid-Kohlenwasserstoff-Fraktion, die durch Extraktion der Hefe *Lodderomyces elongisporus* IMET H 128, gezüchtet auf Erdöldestillat der Siedelage 473—653 K, erhalten wird [4]. Die Abtrennung der Phosphatide erfolgte mit 2,7 Vol.-Teilen Aceton.
Die Kurzwegdestillation wurde in einer Fallfilmverdampferapparatur (Abb. 1) durchgeführt. Zur Destillation wurde die lösungsmittelfreie acetonlösliche Komponente (ALK) des Modellgemisches eingesetzt.

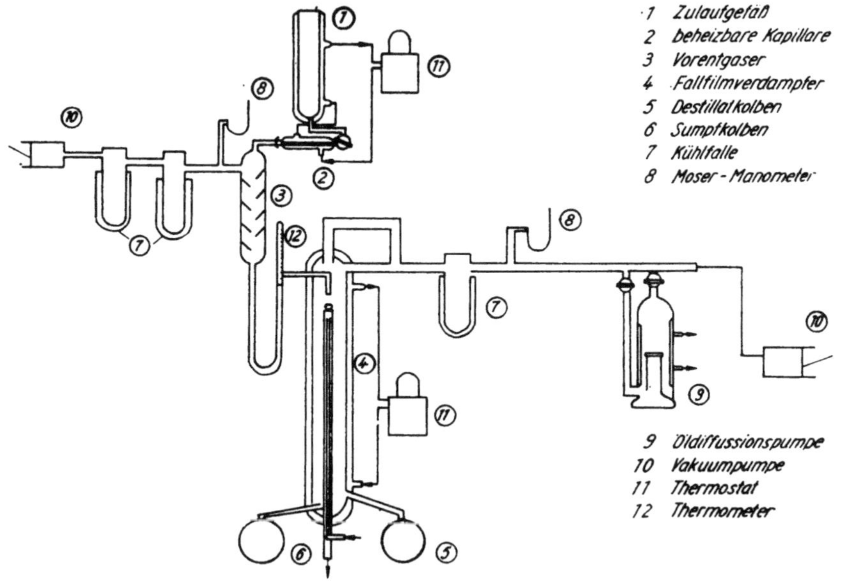

Abb. 1. Fallfilmverdampferapparatur für Kurzwegdestillation

ALK gelangt über eine Kapillare (2), einen Vorentgaser (3) und einen Flüssigkeitsverteiler auf das beheizte Innenrohr des Fallfilmverdampfers (4) und verdampft aus dem Film. Die verdampften Anteile kondensieren an der gekühlten Mantelfläche des Fallfilmverdampfers und werden in der Destillatvorlage (5) gesammelt. Nicht verdampfte Anteile gelangen in die Sumpfvorlage (6). Nicht kondensierbare Dämpfe werden am Kopf des Fallfilmverdampfers abgezogen und kondensieren in der Kühlfalle (7). In der Hauptstufe kann zur Drehschiebervakuumpumpe eine Öldiffusionspumpe zugeschaltet werden. Das mit der Apparatur erreichbare Vakuum liegt bei 2×10^{-3} kPa.
Nach der destillativen Aufkonzentrierung wurde das Sumpfprodukt in Gegenwart eines Antioxidationsmittels mit methanolischer KOH verseift. Durch 4malige Extraktion mit n-Hexan wurde das Unverseifbare (UV) abgetrennt. Aus dem UV wurde nach Abtrennung des Ergosterols (Kristallisation bei 273 bis 282 K) Ubichinon-9 säulenchromatographisch an Silicagel abgetrennt [5].
ALK wurde in drei Stufen aufgearbeitet. Das Sumpfprodukt der Vorstufe wurde in der darauffolgenden Aufarbeitungsstufe als Zulaufgemisch eingesetzt. Die obere Heizmitteltemperatur wurde auf 453 K begrenzt, um die Ergosterolmenge im Destillat gering zu halten.

Ergebnisse

Die Ergebnisse der Kurzwegdestillation sind in Tab. 1—3 zusammengestellt. Sie gestatten folgende Schlußfolgerungen:

— Ubichinone sind bei Temperaturen bis 453 K und einem Druck von $2{,}6 \cdot 10^{-3}$ kPa nicht destillierbar. Sie werden ohne merkliche Zersetzung im Sumpf angereichert.

Tab. 1. Kurzwegdestillation zur Anreicherung von Ubichinon-9

Versuchsbedingungen: 1. Durchlauf Heizmitteltemperatur 433 K, $1{,}3 \cdot 10^{-1}$ kPa
2. Durchlauf Heizmitteltemperatur 433 K, $1{,}3 \cdot 10^{-2}$ kPa
3. Durchlauf Heizmitteltemperatur 453 K, $2{,}6 \cdot 10^{-3}$ kPa

Zusammensetzung von ALK und Sumpfprodukt in Gew.-%

Komponenten	ALK	Sumpfprodukt 3. Durchlauf
Kohlenwasserstoffe	63,4	15,3
Ubichinon-9	0,8	2,7
Ergosterol	1,6	4,2

Tab. 2. Zusammensetzung der Destillate in Gew.-%

Einsatzmange: 1000 g ALK
Destillatmenge des 1. Durchlaufs: 500 g
Destillatmenge des 2. Durchlaufs: 100 g
Destillatmenge des 3. Durchlaufs: 90 g

Komponenten	1. Durchlauf	2. Durchlauf	3. Durchlauf
Kohlenwasserstoffe	99	95	90
Fettsäuren	1	3,6	7,7
Ergosterol	0	1,4	2,3
Ubichinon-9	0	0	0

Tab. 3. Ubichinonanreicherung

Methode	Anreicherung des Ubichinons	ALK	Sumpf	UV
Aufarbeitung von ALK ohne destillative Anreicherung	Gehalt in Gew.-%	0,8		1,3
	Anreicherungsfaktor	—		1,6
Aufarbeitung von ALK unter Einbeziehung einer Kurzwegdestillation (3stufig)	Gehalt in Gew.-%	0,8	2,7	13,5
	Anreicherungsfaktor	—	3,3	16,8

Ausbeute an Ubichinon-9: $> 80\%$ (bezogen auf Gehalt in der Lipid-Kohlenwasserstoff-Fraktion)

Reinheit des isolierten Ubichinon-9: Fp = 314 K (Lit.: 317 K), dünnschichtchromatographisch rein

- Durch die teilweise destillative Entfernung der Kohlenwasserstoffe reichern sich die Ubichinone im Sumpf auf das 2,5 bis 3fache und nach alkalischer Hydrolyse im unverseifbaren Anteil auf das etwa 15fache an.
- Aus den Sumpfprodukten der Kurzwegdestillation lassen sich Ubichinone in einer Ausbeute $> 80\%$ und einer Reinheit $> 95\%$ mit geringerem Aufwand im Vergleich zur Lipid-Kohlenwasserstoff-Fraktion gewinnen.

Eingegangen: 23. 3. 1983

Literatur

[1] CRANE, F. L.: Prog. in Chem. Fats Lipids 7 (1964), 267.
[2] FIECHTER, A. (Ed.): Adv. Biochem. Eng. 17 (1980), 20. Berlin, Heidelberg: Springer-Verlag.
[3] HOLLO, J. u. a.: "The Application of Molecular Distillation", Akadémiai Kiadó, Budapest, 1971.
[4] MÜLLER, H., VOIGT, B.: Acta Biotechnol. 1 (1981), 279.
[5] DDR WP 154448.

Acta Biotechnol. 4 (1984) 3, 297—299

REPORT

Jahrestagung der Biotechnologen im Rahmen der DECHEMA-Jahrestagung 1983 in Frankfurt/Main

Am 15. 6. und 16. 6. 1983 fand erstmals eine Jahrestagung der Biotechnologen im Rahmen der DECHEMA-Jahrestagung 1983 statt. In verschiedenen Vortragsgruppen wurden bestimmte Gebiete der Biotechnologie behandelt.

Vortragsgruppe „Mikrobielle Stoffproduktion“

Ein Übersichtsvortrag über Enzyminhibitoren aus Mikroorganismen (W. Frommer, E. Truscheit, B. Junge, L. Müller, D. D. Schmidt, W. Wingender, Wuppertal) zeigte, daß man zusätzlich zu Antibiotika aus Mikroorganismen weltweit intensiv nach Substanzen mit anderen physiologischen Eigenschaften, z. B. nach Enzyminhibitoren zur Behandlung einer ganzen Anzahl von Krankheiten bei Menschen sucht. Das Problem für eine solche Suche sind neue Tests für bestimmte Stoffgruppen oder für bestimmte physiologische Wirkungen. Für Enzyminhibitoren gibt es bereits neue Suchstrategien.
Mit Hilfe eines Enzymmembranreaktors läßt sich durch Kopplung zweier Enzymsysteme, z. B. der Hexokinase und der Acetatkinase, eine kontinuierliche Regenerierung von ATP für enzymatische Synthesen durchführen (W. Berke, C. Wandrey, Jülich; M. Morr, M.-R. Kula, Braunschweig).
Verschiedene Produkte, wie Tylosin und Nikkomycin (M. Veelken, H. Pape, Münster) sowie Glycerin (B. Bisping, H. J. Rehm, Münster) lassen sich in guten Ausbeuten, die z. T. den Ausbeuten mit freien Zellen entsprechen, z. T. wie bei den Antibiotika höher als die Ausbeuten mit freien Zellen liegen, durch immobilisierte Zellsysteme gewinnen.
Mit *Zymomonas mobilis* kann eine Ethanolfermentation, entkoppelt vom Wachstum der Zellen, mit hohen Ausbeuten durchgeführt werden (S. Bringer, H. Sahm, Jülich). Eine solche Fermentation mit diesem Bakterium ist bereits im halbtechnischen Maßstab in der Erprobung.
Weitere Vorträge über homo- und heterooligomere Enzyme, ausbeutebestimmende Faktoren bei der β-Lactam-Synthese sowie über die Steuerung der Mycelmorphologie bei der Penicillinfermentation zeigten den großen Fortschritt, den die Biotechnologie auch auf diesen Gebieten macht.

Vortragsgruppe „Gentechnik“

Ein halber Tag stand für Übersichtsvorträge aus dem Gebiet der Gentechnik zur Verfügung. Hefen sind zweifellos in Zukunft ausgezeichnete Objekte für die Gentechnik mit Eukaryonten (D. Botstein, Cambridge/USA). Zwar können die Methoden, die an *Escherichia coli* ausgearbeitet worden sind, nicht einfach auf *Saccharomyces cerevisiae* übertragen werden; es wurde jedoch an Beispielen gezeigt, daß beim Gebrauch von Shuttle-Vektoren (z. B. Plasmiden), die sowohl in Bakterien als auch in Hefen vorkommen können, eine erfolgreiche Gentechnik mit Hilfe von Hefen möglich ist. Die gegenwärtigen Ergebnisse über die Manipulation von Genen in Hefen zeigen bereits, daß sich diese Organismen als erfolgversprechende Mikroorganismen für industrielle Produktionen zusätzlich zu den herkömmlichen Verwendungen zeigen.

Die chemische Synthese von DNA ist bereits relativ weit fortgeschritten (E. L. Winnacker, München). Hierbei ist bereits ein Großteil der Methodik automatisiert.

Für die Landwirtschaft hat die Gentechnik möglicherweise in Zukunft eine große Bedeutung (A. Pühler, Bielefeld). Von den verschiedenen Strategien, z. B. für die Einschleusung von neuem genetischen Material in das pflanzliche Genom, wurden einige Beispiele dargestellt. Ein Genvektor für Dikotyledonen und Gymnospermen beruht auf der Tumorbildung durch *Agrobacterium tumefaciens*. Ein DNA-Fragment aus diesem Bakterium (T-DNA) wird in das pflanzliche Genom integriert. Falls Fremdgene in dieser T-DNA inseriert vorliegen, können diese ebenfalls mit in das Pflanzengenom übertragen werden. Transformierte pflanzliche Zellen können auf hormonfreien Medien selektioniert werden, jedoch ist für viele Arten die Regeneration der Gesamtpflanze noch ein Problem.

Die symbiontische Stickstoffixierung mit *Rhizobium*-Arten ist seit langem bekannt. Vielleicht ist es möglich, durch diese Symbiose der Pflanze mit Hilfe veränderter *Rhizobium*-Arten Stoffe zuzuführen, die für sie zum Vorteil sein könnten, z. B. Abwehrstoffe gegen Pflanzenschädlinge.

Mit mitochondrialer DNA, die einer Genmanipulation ausgesetzt worden war, kann bei Eukaryonten eine Leistungsverbesserung erzielt werden, wie dies z. B. bei *Cephalosporium*, einem Pilz, der das Antibiotikum Cephalosporin produziert, der Fall ist (K. Esser, Bochum).

Plasmide aus höheren Pilzen stammen häufig aus Mitochondrien und scheinen auch bei höheren Pflanzen, wie z. B. Mais, Hirse und Bohnen weit verbreitet zu sein. Mitochondrien-DNA, die der der Bakterien sehr ähnlich ist, verspricht eine gute Möglichkeit zur Übertragung in Bakterien oder umgekehrt. Auch für bisher konventionell betriebene Verfahren ist es möglich, die Methode der DNA-Neukombination z. B. nutzbringend zur Entfernung störender Begleitsubstanzen bei der Produktion einzusetzen (R. Mattes, K. Beaucamp, Tutzing).

Das Merkmal der Flockulation von *Saccharomyces cerevisiae* wird durch mindestens zwei Kerngene gesteuert, wobei für die Stärke ihrer Expression extrachromosomale Erbfaktoren verantwortlich sind. Ihre Übertragung in nichtflockulierende Stämme ist möglich und dadurch gelingt eine Verfahrensoptimierung der Ethanolproduktion bei einer bestimmten Prozeßführung (U. Stahl. Bochum).

Bei Hyphenpilzen sind neue Verfahren zur Entwicklung von Klonierungssystemen ausgearbeitet worden (P. Tudzynski, Bochum).

Weitere Vorträge befaßten sich mit dem Einsatz der Gentechnik zur Analyse der genetischen Steuerung der Atmungskette, zur genetischen Analyse N_2-fixierender Mikroorganismen und zur Temperaturstress-regulierten Genexpression bei höheren Pflanzen.

Vortragsgruppe „Fermentation und Aufarbeitung"

Die NADH-Fluoreszenzmessung eignet sich zur Fermentationskontrolle, z. B. bei *Saccharomyces cerevisiae* und *Escherichia coli*. Eine Sauerstofflimitierung wird über diese Messung schneller erkennbar als über die Sauerstoffsondenwerte. Bei der Prozeßsteuerung kann mit dieser Messung hierdurch z. B. auf etwaige Störungen oder Substratmängel reagiert werden (T. Scheper, A. Gebauer, W. Kuhlmann, H.-D. Meyer, K. Schügerl, Hannover).

Die scheinbare Viskosität der Kulturflüssigkeit bei der Penicillinfermentation durch pelletierte und filamentöse Biomasse hat einen nicht unerheblichen Einfluß auf die Peni-

cillinbildung. Mit Hilfe der Messung der Viskosität kann bei der Fermentation zwar keine klare Aussage über den Gehalt an Biomasse gemacht werden, wohl aber über den notwendigen Energieeintrag zur Vermischung der Kulturbrühe (T. LORENZ, R. WITTLER, K. SCHÜGERL, Hannover).
Weitere Vorträge befaßten sich mit Mischzeiten beim Rühren begaster viskoser Flüssigkeiten, mit rechnergekoppelter Datenaufnahme, -Verarbeitung und -Rückgabe bei Fermentationen und mit neuen Verfahren zur Aufarbeitung von Interferon-β und D(-) Lactat Dehydrogenase.

Vortragsgruppe „Mikrobielle Materialzerstörung"

Dieses bisher nur wenig von der Biotechnologie beachtete Teilgebiet hat ein außerordentlich großes Interesse auf der Jahrestagung erfahren. Es hat sich insgesamt gezeigt, daß auf diesem Gebiet durch eine intensivere Bearbeitung als bisher zweifellos viele Mittel eingespart werden können, wenn genaue Kenntnisse über mikrobielle Korrosionen vorliegen und im Anschluß daran solche mikrobielle Korrosionen zielgerecht verhindert werden können.
Bedeutungsvolle mikrobielle Korrosionen finden z. B. bei Kühlschmierstoffen (R. SCHWEISFURTH, Homburg/Saar), in Abwässerkanälen, die aus Beton bestehen (E. BOCK, Hamburg), an Gläsern (W. E. KRUMBEIN, K. JENS, Oldenburg), an Gesteinen (F. E. W. ECKHARDT, Kiel), an Eisen (W. KÖLLE, Hannover), aber auch an Papierprodukten (W. KERNER-GANG, Berlin) statt. Auch Kunststoffe in verschiedener Verwendung können mikrobiell besiedelt werden und dadurch wichtige Eigenschaften verlieren (H. K. FRANK, Karlsruhe).

Vortragsgruppe „Abwasser"

Die Abwasser-Biotechnologie ist, wie viele andere Gebiete, häufig auf der DECHEMA-Jahrestagung behandelt worden. Schon wegen des großen finanziellen Volumens, das zur Beseitigung der Abwässer in Gemeinden und Industrien in der Bundesrepublik Deutschland aufgebracht werden muß, steigt ihre Bedeutung immer mehr an.
In einem Festbett-Umlaufreaktor ist es möglich, mit trägerfixierten Methanbakterien-Essigsäure anaerob zu Biogas abzubauen. Die volumenspezifische Biogasbildung betrug dabei bis zu 45 m^3 pro m^3 und Tag bei einer spezifischen Abbauleistung von 65 kg-CSB pro m^3 und Tag. Die Verweilzeit konnte im kontinuierlich betriebenen Rührkessel bis auf weniger als 8 Stunden reduziert werden (A. *Aivasidis*, C. WANDREY, Jülich).
Die Vor- und Nachteile ein- und zweistufiger Verfahren zur anaeroben Umwandlung proteinhaltiger Industrieabwässer zu Methan und Kohlendioxid wurde experimentell geprüft (W. TRÖSCH, E. STEPHAN, H. CHMIEL, Stuttgart). Abwässer der Zellstoffindustrie lassen sich anaerob im Festbettreaktor mit vielen Vorteilen behandeln. Die CBS-Abbauleistungen liegen bei 90% mit einer gut adaptierten Bakterienpopulation bei einer durchschnittlichen Verweilzeit von 10 Stunden (D. DENGLER, H.-J. KUTZNER, Darmstadt).
Mit Hilfe von *Thiobacillus denitrificans* lassen sich anorganisch belastete Abwässer gut denitrifizieren. Abwässer mit Nitratgehalten bis zu 4 gNO_3/l lassen sich hiermit gut behandeln.
Zusammenfassend läßt sich über die erste Jahrestagung der Biotechnologen sagen, daß hier das breite Spektrum der gesamten Biotechnologie dargestellt wurde. Viele bisher auch unbekannte Aktivitäten in der Bundesrepublik Deutschland wurden präsentiert. Sie zeigen, daß sehr viele Arbeitsgruppen existieren, die mit großem wissenschaftlichen Engagement dieses Gebiet bearbeiten.

Prof. Dr. H. J. Rehm, Institut für Mikrobiologie
Universität Münster, BRD

Acta Biotechnol. 4 (1984) 3, 300

Book Review

H. U. BERGMEYER (Editor in Chief)
J. BERGMEYER, M. GRASSL (Editors)

Methods of Enzymatic Analysis, Vol. I. „Fundamentals“

3rd Edition. Weinheim—Deerfield Beach, Florida—Basel: Verlag Chemie, 1983. 579 S. 240 DM.

„Der Bergmeyer“ als solcher muß nicht besprochen werden, dafür ist er zu bekannt, bewährt und man möchte wohl sagen, „zu unentbehrlich“ in allen biochemischen, klinischen und anderen Laboratorien, die Identifizierungen oder Aktivitätsmessungen von Enzymen und/oder Analysen mit Hilfe von Enzymen durchführen. Durch die rapide Weiterentwicklung und die Einführung neuer Methoden auf dem Gebiet der „enzymatischen Analysen“ seit der letzten Auflage (1974) ergab sich für die Herausgeber die Notwendigkeit, eine völlig neu bearbeitete und sehr stark erweiterte 3. Auflage herauszugeben. Sie wird 10 Bände umfassen, deren 1. Band jetzt vorliegt. Dieser beschreibt in 3 Abschnitten mit insgesamt 21 Kapiteln die für die Praxis relevanten theoretischen und gerätebezogenen Grundlagen der enzymatischen Analyse in einer von den vorhergehenden Auflagen etwas geänderten Reihenfolge der Kapitel. Dem Benutzer werden neben den aus der 2. Auflage bekannten, zumeist aber erweiterten bzw. revidierten Kapiteln zahlreiche neuaufgenommene dargeboten, die moderne Methoden, neue Anwendungen und wichtige Aspekte beinhalten. Zu nennen sind hier: Prinzipien des Enzymimmunoassays, Reflexionsphotometrie, Fluorimetrie, Luminometrie, Potentiometrie, Polarographie, Konduktometrie, Enthalpimetrie, Radiometrie, Turbidimetrie, Neptelometrie, kinetische Eigenschaften von und enzymatische Analysen mit immobilisierten Enzymen, Zuverlässigkeit von Laborergebnissen und Durchführbarkeit der Methoden, allgemeine Hinweise zur Standardisierung bei enzymatischen Analysen, ökonomische Betrachtungen bei enzymatischen Analysen, Vor- und Nachteile verschiedener Meß- und Trennverfahren, Optimierung der Meßbedingungen bei der Bestimmung der katalytischen Aktivität von Enzymproben. Die Kapitel über „Indikatorreaktionen im Sichtbaren“, „Volumenmessungen“ und „Mikrotechniken“ sind sehr stark erweitert.
Am Ende eines jeden Kapitels schließen sich ausführliche Literaturhinweise an, die z. T. Literatur bis 1982 enthalten.
Vergleicht man mit dem 1. Band der 2. Auflage von 1974, so sind im vorliegenden 1. Band der 3. Auflage wegen der gravierenden Erweiterungen nur die Kapitel von A I, A II und A III.2 aus dem 1. Band der 2. Auflage enthalten, während die Kapitel A III.1 und A III.3 sowie die von A IV, A V und alle der Sektion B, die man auch zu den „Fundamentals“ rechnen könnte, offenbar im 2. Band der 3. Auflage enthalten sind.
Papier, Einband und Druck des Buches entsprechen den bekannten Qualitäten des renommierten Verlages.
Von manchem Benutzer könnte als Schönheitsfehler empfunden werden, daß in den Kapiteln über „Absorptionsphotometrie“ und „Mechanisierung und Automatisierung von Analysen“ die Tabellen über kommerzielle Geräte fehlen, ebenso eine geschlossene Darstellung über „fast analyzers“ in einem entsprechenden Unterkapitel. Bei einer gezielten, kurzzeitigen und nur orientierenden Einsichtnahme in die reaktionskinetischen Kapitel wären tabellarische Zusammenfassungen der verschiedenen Gleichungen für verschiedene Mechanismen und Randbedingungen hilfreich.
Diese kritischen Hinweise können das Gesamturteil nicht beeinflussen. Dieser 1. Band der Neuauflage „des Bergmeyers“ bringt in Qualität, Quantität und Aktualität das, was man sich von einem solchen Standardwerk erhofft hat. Gleiches dürfte mit größter Sicherheit auch auf die folgenden Bände zutreffen, auf deren Erscheinen man deshalb mit freudiger Erwartung gespannt sein darf.

R. Berger

Acta Biotechnol. 4 (1984) 3, 301

Book Review

E. FREHLAND

Stochastic Transport Processes in Discrete Biological Systems

Lecture Notes in Biomathematics, Vol. 47
Springer-Verlag, Berlin/Heidelberg/New York, 1982, 169 S.

Die Bedeutung von Membranen für die Regulation biologischer Prozesse rückt immer stärker in das Interesse biophysikalischer Forschung. Es hat sich gezeigt, daß stochastische Modelle, die auf diskreten Zustandsräumen beruhen, die Eigenschaften solcher Transportprozesse an biologischen Membranen besser beschreiben als deterministische stetige Modelle.
Eckart FREHLAND versteht nun sein Buch, das aus einer Vorlesungsreihe über die Anwendung stochastischer Prozesse für fortgeschrittene Studenten hervorgegangen ist, als Einführung zu einer einheitlichen Theorie der Fluktuationen in komplexen biologischen Transportsystemen. Es sind zum wesentlichen Teil eigene neuere Ergebnisse verarbeitet worden.
Der Bogen spannt sich von einer Einführung zu stochastischen Prozessen über Fluktuationen an Fließgleichgewichtszuständen bis zu Fluktuationen bei Übergangsprozessen. Das theoretische Konzept ist klar aufbereitet und den Möglichkeiten entsprechend vollständig abgearbeitet. Aus der Zielstellung folgt, daß bis auf ein Beispiel kein Vergleich der Modelle mit praktischen Ergebnissen erfolgt. Deshalb wird sich dieses Buch auch kaum als Brücke von der Theorie zu experimentell arbeitenden Biologen oder Biochemikern eignen. Es ist für die Theoretiker sehr zu empfehlen, die sich intensiver mit der Problematik stochastischer Prozesse an biologischen Membranen beschäftigen möchten.
Da auch für bestimmte biotechnologische Verfahren Transportprozesse durch Membranen von entscheidender Bedeutung sind, sollte geprüft werden, inwieweit stochastische Einflüsse bei der Modellierung der Verfahren Berücksichtigung finden müssen.

Th. Bley

Acta Biotechnol. 4 (1984) 3, 302

Book Review

H. G. Gassen, A. Lang

Chemical and Enzymatic Synthesis of Gene Fragments
A Laboratory Manual

Weinheim—Deerfield Beach, Florida—Basel. Verlag Chemie 1982,
18 Beiträge, Index, Liste der Ausrüstungen, 69 Zeichnungen und Abbildungen

Sequenzspezifische DNA-Abschnitte werden seit Mitte der 70er Jahre in immer größerem Umfang bei allen Grundoperationen der Gentechnik benötigt. Die kombinierte chemisch-enzymatische Synthese von DNA-Fragmenten wurde noch bis Ende der 70er Jahre nur in einigen hochspezialisierten Laboratorien durchgeführt. Die Weiterentwicklung der chemischen Synthesetechniken, besonders des Triester- und Phosphit-Triesterverfahrens und ihre Adaptierung an feste Phase hat in den letzten Jahren zu einem enormen Aufschwung der chemischen d-Oligonucleotid- und damit auch der Gen-Synthese geführt, so daß gegenwärtig derartige Techniken quasi als Standardmethoden Einzug in die Laboratorien der Gen-Technologen und Molekularbiologen und der Industrie halten. H. G. Gassen und A. Lang haben Laborprotokolle führender Gruppen für einen Kurs „Automatische chemische und enzymatische Gen-Synthese“ zusammengefaßt. Zusammen mit den Originalarbeiten setzt diese Anleitung Wissenschaftler, die mit diesem Gebiet nicht so vertraut sind, in die Lage, routinemäßig DNA-Sequenzen zu synthetisieren.
Unter vier Hauptthemen werden die jeweiligen grundlegenden Methoden in sorgfältig aufbereiteten Laborprotokollen dargeboten.
Jedem Beitrag sind Zitate wichtiger Original- und Übersichtsarbeiten beigefügt.

1. Chemische Synthese sequenzspezifischer Polydesoxyribonucleotide: Triestersynthese an Kieselgur-Polyamid-Träger (Gait), Polyamid-Träger (Caruthers; Seliger et al., Winnaker und Perper), automatische sowie Festphasensynthese (Cook et al.), neue Darstellungsverfahren geschützter Mono- und Dimere (Fritz et al., Van Boom et al.), präparative Gewinnung von dOligonukleotiden aus chemisch abgebauter DNA (Schott). Es fehlen Laborprotokolle der Gruppen um Itakura, Ogilvie und Kempe.
2. Biologische Modifikation von Gen-Fragmenten: In diesem Abschnitt werden Grundlagen der Gentechnik kurz referiert (Schmitt und Gassen; Klock; Gatz und Hillen), die allerdings im Manual "Molecular Cloning“ von Maniatis et al. (Cold Spring Harbor, 1982) gründlicher und umfassender dargestellt sind, und eine Strategie zur Produktion größerer Mengen von DNA-Sequenzen für physiko-chemische Methoden (Hillen) vorgestellt.
3. Enzymatische Synthese sequenzspezifischer RNA-Abschnitte: Mittels Polynucleotid-Phosphorylase (Lang und Gassen) und RNA-Ligase (Uhlenbeck; Ohstuka; Eckert). Verfahren zur chemischen Synthese sind nicht aufgenommen.
4. Reinigung und Charakterisierung von DNA- und RNA-Sequenzen: Im Vordergrund stehen die schon klassisch zu nennenden Reinigungsverfahren mittels HPLC im analytischen bis präparativen Maßstab (Laughlin und Krusche); Vorstellung eines umfassenden Algorithmus einschließlich Bestimmung der Basenzusammensetzung und Sequenzanalyse nach Maxam-Gilbert von Fritz et al.; zweidimensionale Fingerprinttechnik (Frank und Blöcker).

Dieses wertvolle Praktikumsbuch ist als Arbeitsanleitung Chemikern, Biochemikern und Molekularbiologen zu empfehlen, die auf dem Gebiet der chemisch-enzymatischen Synthese und Charakterisierung von DNA- und RNA-Abschnitten arbeiten.

S. Rosenthal

Acta Biotechnol. 4 (1984) 3, 303

Book Review

I. CHIBATA, S. FUKUI, L. B. WINGARD jr.

Enzyme Engineering Vol. 6

New York, London: Plenum Press 1982, X, 538 S., 50.50 $

Der Band enthält den Text der 116 Referate auf der 6. Enzyme-Engineering Conference vom 20.—25. September 1981 in Kashakojima, Japan.

Nach einer Übersicht über die Geschichte dieser aller zwei Jahre in verschiedenen Orten stattfindenden Konferenz von L. B. WINGARD jr. folgt von H. UMEZAWA „Recent Studies on Antibiotics and Low Molecular Weight Enzyme Inhibitors", die neueren Ergebnisse dieser Arbeitsgruppe im guten Überblick.

Im folgenden Text sind die Vorträge in 8 Gruppen eingeteilt:

— Large Scale Production and Purification of Biomolecules
— Application of Biocatalysts for New Reactions and Organic Synthesis
— New Immobilization Techniques of Biomaterials and Their Application
— Industrial Application of Immobilized Biomaterials
— Biomass Conversion with Energy Production
— Analytical Application of Immobilized Biomaterials
— Medical Applications of Enzyme Technology
— Genetic Engineering for Enzyme (or Important Biological Substances) Products.

Am Schluß wird von E. KATCHALSKI-KATZIR ein Gesamtüberblick über neue Entwicklungen und über Zukunftsaussichten gebracht.

Der Band gibt einen durchaus repräsentativen Querschnitt über die Aktivitäten auf dem Gebiet des technologisch orientierten Einsatzes von freien und vorzugsweise immobilisierten Enzymen einschließlich von integrierten biokatalytischen Systemen, d. h. Zellen, vorwiegend mikrobiellen Ursprungs, etwa nach dem Stand zu Beginn des Jahres 1981.

So ingeniös und anregend viele der angebotenen Lösungen sind, sie werden sich den harten ökonomischen und technologischen Bedingungen der Praxis zu stellen haben und nicht in allen Fällen überstehen. Auch die Hilfe, die man sich vom „genetic engineering" für das Enzyme-Engineering erhoffen kann, wird nicht alle Schwierigkeiten beseitigen und vielleicht einige neue mit sich bringen.

Bei aller kritischen Haltung, die auch in einigen Beiträgen sehr realistische Einschätzung veranlaßt, bringen die Artikel dieses Bandes doch auch ein Bild der großen in dieser Technologie steckenden Möglichkeiten.

Man kann den Sätzen des Schlußwortes nur zustimmen, in denen die Voraussetzungen für weitere Fortschritte aufgezählt werden: Billige Methoden für Produktion und Reinigung von Biokatalysatoren in großem Maßstab; Neue Methoden der Enzym- und Zellstabilisierung; Verbesserte Immobilisierungsmethoden; Methoden der Cofaktorregenerierung; Ausdehnung der Anwendungen in Industrie und organischer Synthese; Wirksamere Methoden der Biomasseumwandlung und Energieerzeugung; Entwicklung neuer analytischer Methoden.

Alles in allem eine instruktive und anregende Vortragssammlung, die in allen einschlägig arbeitenden Einrichtungen zur Verfügung stehen sollte.

G. Langhammer

Acta Biotechnol. 4 (1984) 3, 304

Book Review

F. K. Fong

Light Reaction Path of Photosynthesis

Molecular Biology, Biochemistry and Biophysics 35

Berlin, Heidelberg, New York: Springer-Verlag 1982. 138 DM; 55.20 $

Die vorliegende Monographie aus der Feder neun namhafter Wissenschaftler bietet einen Überblick über die Entwicklung der Photosyntheseforschung der letzten zehn Jahre. Behandelt werden die Einzelschritte des Photosyntheseprozesses in der Reihenfolge ihres Auftretens.
Die dargelegten experimentellen Methoden und Ergebnisse sind repräsentativ für den Fortschritt der instrumentellen Techniken und die konzeptionelle Entwicklung auf diesem Gebiet im betrachteten Zeitraum. Besonders Pikosekunden-, zeitaufgelöste und magnetische Resonanzspektroskopie haben gemeinsam mit strukturellen und photoelektrochemischen Photosynthesemodellen Hinweise auf die molekularen Mechanismen des Energieübergangs von den Antennensystemen auf die Reaktionszentren und der schließlich zur CO_2-Reduktion führenden weiteren Teilschritte erbracht. Komplizierte physiologische Zusammenhänge werden auf der Basis moderner photochemischer Erkenntnisse erklärt. In mehreren Bereichen kommt es zu einer Gegenüberstellung von in vivo- und in vitro-Ergebnissen. Viele der vorgelegten Befunde sind neuartig. Von besonderem Interesse sollte die Erkenntnis sein, daß die photosynthetische CO_2-Reduktion ein integraler Bestandteil der Lichtreaktion des Chlorophylls ist.
Das vorliegende Buch mit seinen 400 Literaturzitaten sollte aus den genannten Gründen für den in der Photosynthese-Grundlagenforschung tätigen Wissenschaftler eine nahezu unersetzliche Informationsquelle sein.

R. Roth

Acta Biotechnologica

Volume 4 1984 Number 3

Contents

Acta Biotechnologica is indexed in Current Contents/ET & AT;
Chemical Abstracts; Biotechnology Abstracts; Excerpta Medica

Bei Fragen zur Produktsicherheit wenden Sie sich bitte an:
If you have any questions regarding product safety,
please contact:

Walter de Gruyter GmbH
Genthiner Straße 13
10785 Berlin
productsafety@degruyterbrill.com